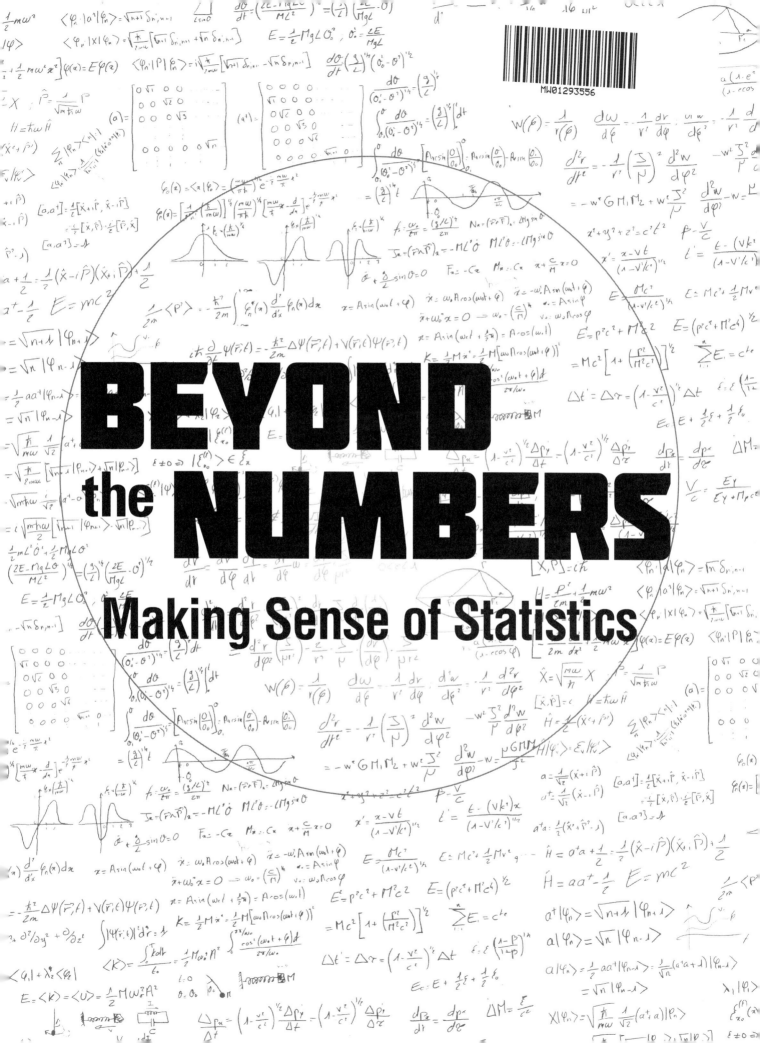

BEYOND the NUMBERS

Making Sense of Statistics

BEYOND the NUMBERS
Making Sense of Statistics

Edwin P. Christmann

National Science Teachers Association

Arlington, Virginia

National Science Teachers Association

Claire Reinburg, Director
Jennifer Horak, Managing Editor
Andrew Cooke, Senior Editor
Wendy Rubin, Associate Editor
Agnes Bannigan, Associate Editor
Amy America, Book Acquisitions Coordinator

Art and Design
Will Thomas Jr., Director, Interior Design
Lucio Bracamontes, Graphic Designer, cover and interior design

Printing and Production
Catherine Lorrain, Director
Nguyet Tran, Assistant Production Manager

National Science Teachers Association
Francis Q. Eberle, PhD, Executive Director
David Beacom, Publisher

Copyright © 2012 by the National Science Teachers Association.
All rights reserved. Printed in the United States of America.
15 14 13 12   4 3 2 1

Library of Congress Cataloging-in-Publication Data

Christmann, Edwin P., 1966-
  Beyond the numbers : making sense of statistics / by Edwin P. Christmann.
    p. cm.
  Includes index.
  ISBN 978-1-935155-25-6
  1. Mathematical statistics. I. Title.
  QA276.12.C468 2012
  519.5--dc23
                                    2011033960

eISBN 978-1-936959-92-1

   NSTA is committed to publishing material that promotes the best in inquiry-based science education. However, conditions of actual use may vary, and the safety procedures and practices described in this book are intended to serve only as a guide. Additional precautionary measures may be required. NSTA and the authors do not warrant or represent that the procedures and practices in this book meet any safety code or standard of federal, state, or local regulations. NSTA and the authors disclaim any liability for personal injury or damage to property arising out of or relating to the use of this book, including any of the recommendations, instructions, or materials contained therein.

Permissions
Book purchasers may photocopy, print, or e-mail up to five copies of an NSTA book chapter for personal use only; this does not include display or promotional use. Elementary, middle, and high school teachers may reproduce forms, sample documents, and single NSTA book chapters needed for classroom or noncommercial, professional-development use only. E-book buyers may download files to multiple personal devices but are prohibited from posting the files to third-party servers or websites, or from passing files to non-buyers. For additional permission to photocopy or use material electronically from this NSTA Press book, please contact the Copyright Clearance Center (CCC) (*www.copyright.com*; 978-750-8400). Please access *www.nsta.org/permissions* for further information about NSTA's rights and permissions policies.

CONTENTS

vii   **PREFACE**

1   **CHAPTER 1**
INTRODUCTION

9   **CHAPTER 2**
SCALES AND NUMBER DISTRIBUTION

27   **CHAPTER 3**
CENTRAL TENDENCY AND VARIABILITY

49   **CHAPTER 4**
STANDARD SCORES

69   **CHAPTER 5**
INTRODUCTION TO HYPOTHESIS TESTING: THE Z-TEST

93   **CHAPTER 6**
THE T-TEST

141   **CHAPTER 7**
ANALYSIS OF VARIANCE (ANOVA)

165   **CHAPTER 8**
CORRELATION

183   **CHAPTER 9**
THE CHI-SQUARE (X^2) TEST

207   **APPENDIX**
AREAS OF THE STANDARD NORMAL DISTRIBUTION

209   **INDEX**

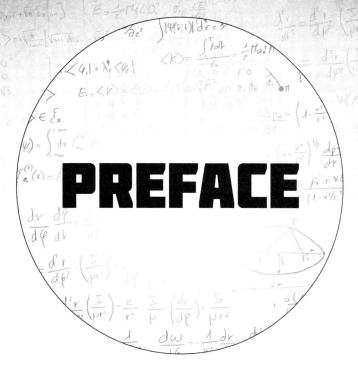

PREFACE

This book is designed for use with introductory statistics courses, educational assessment courses, science and mathematics educations courses, and in-service teacher development. It presents a nonthreatening, practical approach to statistics with the template for applications of the graphing calculator. It is my hope that this book is a resource that gives educational planners the means to examine prescribed methodologies for data analysis to make informed decisions.

We begin with an overview of the development and current purposes of statistics and some of the individuals who have played significant roles in this movement that spans more than 2,000 years. Then statistics is introduced in a mathematically friendly manner that presents a variety of ways to calculate data. In addition, the book shows readers how to make computations by hand and with a calculator.

This book provides step-by-step instructions for understanding and implementing the essential components of statistics. Its numerous examples and their accompanying explanations will serve as models for devising methods of statistics that are commensurate with data driven decision making.

Chapter 1 presents an overview of the historical development and current purposes of statistics and some of the individuals who have played significant roles in this 2,000-year-old movement.

Chapter 2 demonstrates how to organize your data. Also, you are introduced to scales of measurement, the normal curve, and skewed distributions of data.

Chapter 3 familiarizes you with the concepts of central tendency (i.e., mean, median, and mode) and variation (i.e., standard deviation and variance). Knowledge of these basic statistical concepts, which you will apply to data, will provide a foundation for all other statistics concepts.

Chapter 4 is an orientation to the procedures that enable you to determine class rank, percentile rank, and standard scores from either classroom or norm-referenced data. This chapter also provides you with an understanding of achievement and aptitude test data.

Chapter 5 presents an overview of the historical development hypothesis testing and introduces the z-test.

Chapter 6 demonstrates how to compute t-tests. Subsequently, examples of the one-sample t-test, the independent t-test, and the dependent t-tests are introduced.

Chapter 7 not only explains the importance of Analysis of Variance (ANOVA), but it also gives you step-by-step instructions for its calculation.

Chapter 8 introduces you to the concept of correlation by demonstrating the different assessment instruments.

Chapter 9 demonstrates how to compute the chi-square goodness of fit and independence tests. Subsequently, nonparametric statistical methods are introduced and examples are explored.

SUPPLEMENT

As a supplement to the text, a website that offers chapter summaries, tutorials, and PowerPoint presentations will assist in the organization and presentation of the material [*http://srufaculty.sru.edu/edwin.christmann/epc2.htm*].

ACKNOWLEDGMENTS

In this life, anything can be made possible with help and support. Thanks to my wife, Roxanne, and my three children, Lauren, Forrest, and Alexandra, I was given that help and support. In addition, the NSTA staff has given me the opportunity to complete this project, and I thank them for their support and guidance. A teacher once told me that everything in life is done for a single purpose. He said it can be found in the expression, *Ad Majorem Dei Gloriam*. My hope is that it was and it will always be for that single purpose. Thank you!

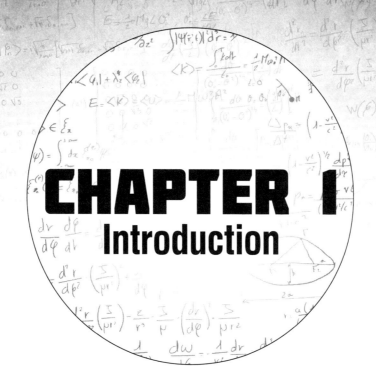

CHAPTER 1
Introduction

OBJECTIVES

When you complete this chapter, you should be able to
1. demonstrate an understanding of the history of statistics,
2. list the important early statisticians,
3. compare and contrast different statisticians,
4. demonstrate an understanding of the development of ANOVA, and
5. explain how advances in technologies have improved statistical analysis.

Key Terms

When you complete this chapter, you should understand of the following terms:
    statistics
    statistician
    probability
    inferences
    correlation
    regression to the mean
    ANOVA
    statistical calculations

"While the individual man is an insoluble puzzle, in the aggregate he becomes a mathematical certainty. You can, for example, never foretell what any one man will be up to, but you can say with precision what an average number will be up to. Individuals vary, but percentages remain constant. So says the statistician."

—Sir Arthur Conan Doyle

CHAPTER 1

INTRODUCTION

In about 1749, a German political geographer named Gottfried Achenwall coined the term *statistik*, which was used as a synonym for "state arithmetic." As a geographer, he used the term to describe the strengths and weaknesses of a nation state. The term *statistics* today, however, is used to describe a methodology for drawing conclusions based on observations (Tankard 1984). More specifically, statistics is a branch of mathematics and science that relates to the collection, analysis, interpretation, and presentation of data. In essence, statistics involves the gathering of data and the application of a statistical test so that a researcher can draw conclusions.

A BRIEF HISTORY OF STATISTICS

In all likelihood, prehistoric people began to bring order to our world by counting. My guess is that by using their fingers and their toes, numbers became useful tools for decision making. Over time, however, more complex mathematical procedures were developed by mathematicians such as Euclid, Pythagoras, and Archimedes in the magnificent cultural centers of Greece and Alexandria. However, after the Hellenistic Age (336 BC to 30 BC), as a result of library collections being displaced during wars, mathematics found its way to distant places like Arabia and India (David 1962).

When Rome fell circa 455 AD, the church became the bastion of learning and scholarship throughout the western world. Subsequently, mathematics during the Middle Ages (circa 500 AD to 1500 AD), focused on the applications of astronomy and theology. The Venerable Bede—a monk, author, scholar, and historian—invented a system of counting and published on arithmetic, while Alcuin of York, along with Charlemagne, advocated mathematics by establishing universities throughout Europe. However, the development of statistics seems to have been stalled by the fact that probability, a component of statistics, was associated with gaming. Since gaming is considered a vice prevalent among nonbelievers, the study of probability was not emphasized.

Ironically, circa 1000 AD, a book that could have been used to further develop mathematics—and possibly could have unlocked answers to many mathematical and statistical questions—was hidden in a monastery library in Bethlehem. We now know that Archimedes' manuscript, *The Method*, written circa 250 BC, was the first identified application of calculus to solve area problems. This method could have been useful in computing probabilities. However, since Archimedes' book was lost, it was not until the 15th century when Isaac Newton and Gottfried Leibniz developed calculus as a methodological tool to explain phenomena. Consequently, the applications of mathematics to human innovations hastened the development of technologies that would eventually lead to lunar landings and microcomputers.

SIR FRANCIS GALTON

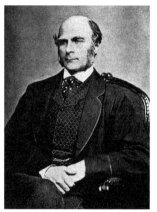

As the scientific community began to make advances, researchers relied on their results to predict and forecast their findings. Hence, so that it could be generalized to populations, a tool was needed for the applications of quantitative data. Over time, the development of statistical tests gave researchers the opportunity to make *inferences*—deductions from sample data that could explain what the population might think. For example, the English scientist, Sir Francis Galton (1822–1911), used statistical analysis to classify and identify fingerprints. However, it was his investigation of intelligence and inherited traits that led to his discovery of "regression to the mean." *Regression to the mean* is a phenomenon in which a variable that is extreme on its first measurement will tend to be closer to the center of the distribution on a later measurement. Consequently, Galton coined the term *correlation* and was able to create a method for inferring functional relations from correlations.

KARL PEARSON

Galton's discovery influenced the work of English mathematician Karl Pearson (1857–1936), which led to Pearson's development of linear regression and the Pearson correlation. In essence, correlation indicates the strength and direction of a linear relationship between two random variables. For example, there is a positive correlation between cigarette smoking and the incidence of lung cancer; there is a negative correlation between the hours-per-week watching television and GPA. Furthering Galton's research, Pearson used Galton's data sets to examine the regression of sons' heights upon that of their fathers. Subsequently, to illustrate correlation, Pearson built a three-dimensional representation of this data set, which is used to model correlation today. Albert Einstein was influenced by Pearson's book *The Grammar of Science*, which covered several topics Einstein later explored. For example, Pearson asserted that the laws of nature are relative to perceptive ability, and he speculated that an observer who traveled faster than light would see time reversal, similar to a cinema film being run backwards. In addition, Pearson's writings mentioned topics such as antimatter and the fourth dimension.

CHAPTER 1

WILLIAM SEALY GOSSET

Under the tutelage of Karl Pearson, a young Oxford alumnus named William Sealy Gosset (1876–1937) began his work as a statistician. However, he first gained notoriety when he was an employee of the Guinness Brewery. Gossett did research on small sample sizes as they related to the ingredients in Guinness beer. Subsequently, he was able to improve the selection of barley by developing the *t-distribution* and *the probable error of the mean*. Ironically, since the Guinness Brewery had a policy to protect trade secrets, its employees were not permitted to publish any research findings. Cleverly, William Gosset published his findings in the prestigious journal of Karl Pearson, *Biometrika*, under the pseudonym *Student*. Although Karl Pearson did not seem moved by Gosset's findings, the statistician Sir Ronald Aylmer Fisher (1890–1962) found Gosset's manuscript most useful by using it derive the *t-statistic* and his theory on *degrees of freedom* from Gosset's work.

SIR RONALD FISHER

Sir Ronald Fisher (1890–1962) was one of the pioneers of the field of statistics. Through his studies in crop variation, Fisher created the modern methodologies for the design of experiments and the Analysis of Variance (ANOVA). Some believe that it was Fisher who shaped the fundamentals of modern statistics. Undoubtedly he did; however, it may have been his terrible eyesight that forced him to assimilate an unconventional means to conceive images of calculations in his mind without having to use his eyes for reading and writing his computations. Perhaps his unique aptitude for mathematics found him a niche at the Rothamsted Agricultural Experiment Station, where he developed statistical procedures for experimental designs with fertilizers.

Fisher's development of the ANOVA concept was possibly his supreme accomplishment. Today, the ANOVA is paramount to advances in statistics and research in that the ANOVA provides multiple concurrent answers to research questions. Fisher's principal idea was to arrange an experiment with different levels of variables that are different in terms of the factors or treatments applied in them. The different outcomes could then be analyzed through a comparison of different factors or combinations of factors. Fisher's monumental work led to multivariate analysis, post-hoc tests, and solutions to statistical bias.

Fisher spent the last years of his life researching and debating the conclusion that smoking causes lung cancer. Fisher argued that the correlation does not establish cause-and-effect and that causation could not be established. Ironically, Sir Ronald Fisher, an avid smoker, died of cancer in 1962.

SUMMARY

Statistics is generally regarded as a subfield of mathematics; however, many consider it a related discipline. Many universities maintain separate mathematics and statistics departments. Nonetheless, statistics is also taught in departments such as education, psychology, and engineering. The advent of technology via mainframe computers has made it possible for scientists and mathematicians to perform complex calculations at much faster rates. However, as technology improves over time, computers have transformed into smaller, handheld devices. Hence, calculators became accessible to students as portable and affordable problem-solving tools. Subsequently, the microcomputer has evolved into a tool that has become a mechanism to bolster statistical applications.

Statistics is an academic discipline that is closely related to mathematics and provides students with a foundation for conducting research. The presentation of statistics to students, however, is often methodologically obscure, which prohibits the application of problem-solving skills. Perhaps this is because the mathematical core of statistical analysis is not congruent with the aptitudes of most students enrolled in statistics courses. If this is the case, current technologies (e,g., graphing calculators and statistics packages such as SPSS) are considered to be beneficial in assisting students with complex statistical calculations. These technologies have led to more accurate instruments for the application of statistics and have made possible substantial enhancements in statistical practices.

Today the use of statistics has broadened far beyond its origins. Individuals and organizations use statistics to understand data and make informed decisions throughout the natural and social sciences, medicine, business, and other areas. Hopefully, this book will give you the information that you need to understand data analysis and help you to make decisions that are inquiry-based.

INTERNET RESOURCES

- The site of the PBS television series NOVA explains the history of Archimedes' manuscript *The Method*, revealing the mind of the Greek genius. *www.pbs.org/wgbh/nova/archimedes/palimpsest.html*
- This video instructional series on statistics for college and high school classrooms and adult learners contains 26 half-hour video programs. *www.learner.org/resources/series65.html*
- This site is supported by UCLA and provides images and biographies of statisticians. *www.stat.ucla.edu/history*

- This site is supported by the Federal Government and makes data available from more than 100 agencies. *www.fedstats.gov*

CHAPTER REVIEW QUESTIONS

1. Statistics involves the gathering of data and the application of a statistical test so that
   a. data specifications can be framed.
   b. comprehensive testing can be applied to data.
   c. researcher can draw conclusions.
   d. standards for evaluation can be established.

2. After the Hellenistic Age (336 BC to 30 BC), mathematics found its way to
   a. Rome.
   b. Greece.
   c. England.
   d. Arabia and India.

3. His investigation of intelligence and inherited traits led to his discovery of "regression to the mean."
   a. Sir Francis Galton
   b. Karl Pearson
   c. Sir Isaac Newton
   d. Albert Einstein

4. The ANOVA can be used to analyze the different outcomes by comparing
   a. data.
   b. combinations of factors and/or different factors.
   c. the probable error of the mean.
   d. knowledge, attitudes, and skills.

5. This person published his findings in the prestigious journal <u>Biometrika</u>, under the pseudonym *Student*.
   a. Sir Ronald Fisher
   b. William Sealy Gosset
   c. Karl Pearson
   d. Sir Francis Galton

ANSWERS: CHAPTER REVIEW QUESTIONS

1. C. Researchers can draw conclusions.
2. D. Arabia and India.
3. A. Sir Francis Galton
4. B. combinations of factors and/or different factors.
5. B. William Sealy Gosset

REFERENCES

David, F. N. 1962. *Games, gods, and gambling*. Charles Griffin: London.

Fisher, R. A. 1921. Studies in crop variation. *Journal of Agricultural Science* 11: 237–265.

Fisher, R. A. 1956. *Statistical methods and scientific research*. Edinburgh: Oliver and Boyd.

Pearson, K. 2007. *The grammar of science*. Cosimo Classics: New York, NY.

Salsburg, D. 200). *The lady tasting tea: How statistics revolutionized science in the twentieth century*. Henry Holt: New York, NY.

Tankard, J. W. 198) *The statistical pioneers*. Schenkman: Cambridge, MA.

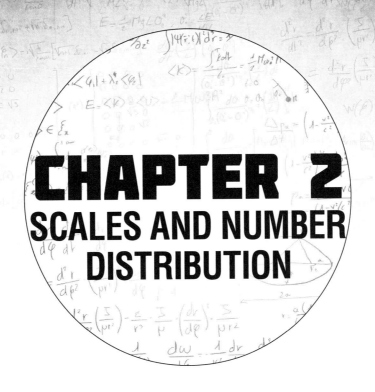

CHAPTER 2
SCALES AND NUMBER DISTRIBUTION

OBJECTIVES

When you complete this chapter, you should be able to
1. demonstrate an understanding of the different scales of measurement and how they relate to educational measurement;
2. compare and contrast the different types of data as they relate to educational measurement;
3. select, create, and use appropriate graphical representations of data, including histograms;
4. represent data using tables and graphs such as line plots, bar graphs, and line graphs;
5. recognize and generate equivalent forms of decimals through rounding; and
6. describe the shape and important features of a set of data and compare related data sets, with an emphasis on how the data are distributed in school settings.

Key Terms

When you complete this chapter, you should understand the following terms:

array	histogram
bar graph	interval scale
bell curve	nominal scale
Cartesian grid	normal curve
continuous variable	normal distribution
discrete variable	ordinal scale
distribution	ratio scale
frequency distribution	rounding numbers
frequency graph	simple bar graph
frequency polygon	skewed distributions
frequency table	variables

CHAPTER 2

Statistics is a branch of science that deals with the collection, analysis, interpretation, and presentation of numerical data. We use statistics every day to solve problems. Geneticist Gregor Mendel, for example, experimented with pea plants, which led to the development of theories of dominant and recessive genes. Similarly, teachers use statistics to average classroom grades to determine what students have learned as a result of classroom instruction. In this chapter, we will demonstrate how you can use the powerful tool of descriptive statistics in your classroom to help you summarize, organize, and simplify data.

To use statistics, we must first take measurements. According to Allen and Yen (2002), measurement is "the assigning of numbers to individuals in a systematic way as a means of representing properties of the individuals." For example, when you look at a thermometer and note that the outside temperature is 67°F, you have measured the temperature, but what does this measurement mean? In itself, it's just a number. When we arrange measurements into scales, however, we can begin to understand and to compare them. This is what teachers do with statistics when taking measurements of student performance, both formative—ongoing measurements—and summative—end-of-the-semester measurements. That is, teachers measure student performance on a continual—formative—and on a final—summative—basis.

As you already may have found, test scores are sometimes not easy to interpret. For example, if a student scored a 610 on the mathematics section of the SAT, you might not be able to interpret exactly what that score represented. If a test were based on a measurement scale that ranged from a minimum low score of 0 to a maximum high score of 100, however, you might find the result more familiar. As an analogy, the temperature of 72°F is the equivalent of 22.22°C. Most Americans are more familiar with measuring temperature with the Fahrenheit scale; the Celsius system can be confusing unless you are a scientist who uses the Celsius scale in your work. In the field of education, we work with numerical measurements, reported as scores that are assigned to one of four types of measurement scales—*ordinal, nominal, interval,* or *ratio*.

MEASUREMENT SCALES

Ordinal Scales. *Ordinal scales* classify numbers into a set in which each number is either less than, equal to, or greater than every other number. Ordinal scales imply an ordered number sequence. For example, a teacher may take height measurements and arrange students in order from the tallest to the shortest student. In a similar fashion, the marks on examinations can be put in order from the lowest score to the highest score. It is important that, as a teacher, you understand how to interpret data, because misinterpretation or misuse of statistics can be misleading. For example, if you were to rank order the members of a professional basketball team from tallest to shortest, the results should not be interpreted to mean that the shortest player at 6'5" is "short." Such an assumption would be misleading. Similarly, a student who has the lowest math score in a gifted class is not necessarily a poor math student.

Nominal Scales. *Nominal scales* classify numbers into different (named) categories. For example, a teacher might use a nominal scale to assign students into two different groups on the basis of gender. On this basis, the teacher might observe that, of 26 students, 12 male students scored 80% or higher on an exam, while 14 female students scored 80% or higher on the same examination.

Interval Scales. *Interval scales* classify numbers in equal units, but they do not have a true zero. Temperature is measured on an interval scale: A difference of 1 degree is always the same size. Most psychological tests use interval scales, and, because the scores represent equal units, they can be added or subtracted. Thus, a student whose IQ is 120 can be said to have an IQ that is 20 points higher than one whose IQ is 100. Because IQ measurement does not have a true zero, however, we cannot say that an IQ of 120 is twice as high as an IQ of 60. In actual mental capability, a score of 120 is probably much more than twice a score of 60.

> ### Why Is Freezing 32°F?
> My favorite example of the interval scale is the Fahrenheit temperature scale. Have you ever wondered how Gabriel Fahrenheit arrived at 32° as the freezing point and 212° as the boiling point of water? Hint: Subtract the difference, and you get 180°, which in simple geometry, gives you the diameter of a circle (remember a circle has 360°). The point here is that Gabriel Fahrenheit selected his scale, which is an interval scale, arbitrarily.

Ratio Scales. The *ratio scale,* like the interval scale, is a classification of numbers expressed in equal units. However, unlike the interval scale, the ratio scale has a true zero. Some examples of the ratio scale are units of time, distance measurements on a ruler in inches, and the measurement of mass in grams, all of which are typically used in the physical sciences. All of these numbers can be multiplied and divided, as well as added and subtracted. A ratio scale is distinguished from an interval scale by the fact that it has a true zero (as in money), while an interval scale does not have a true zero (as in intelligence). Ratio scales, however, are rarely encountered in educational statistics, testing, and measurements.

VARIABLES

Variables are symbols that can take on a variety of numerical values. In statistics, many of the variables that we use are displayed either italicized or as Greek letters. In educational testing and measurement we use these variables to represent scores. For example, if x represents a set of test scores, x_1 is the first examinee's score, x_2 the second examinee's score, x_3 the third examinee's score, and so forth (see Table 2.1, p. 12). In addition, there are two other types of variables, *discrete* and *continuous,* that relate to educational testing.

Continuous Variables. A *continuous variable* is a variable that can be divided into an

CHAPTER 2

Table 2.1

	X
X_1	87
X_2	93
X_3	65
X_4	78
X_5	87
X_6	99
X_7	75
X_8	68
X_9	65
X_{10}	88

Table 2.2

Science Test Results

77
77
67
65
97
90
89
89
100
98
64
50
42
74
73
72
72
70
89
81
80
80
79

infinite number of fractional parts. For example, time is a continuous variable because an infinite number of possible values falls between any two observations (see Figure 2.1.). In education, we most often work with continuous variables from the interval measurement scales.

Discrete Variables. A *discrete variable* cannot be divided into fractional parts. Some examples are class size, family size, gender, or the combined value of rolling dice. Gender, for example, observes males and females, two values that cannot be divided into fractional parts. Thus, gender is a discrete variable.

IQs, however, are numerical values that arbitrarily range from 25 and below, indicating profound mental retardation, to 130 and above, which indicates a very superior intelligence. Unlike the example of time, however, IQs increase by increments of one unit (i.e., 100, 101, 102, and so forth) along a cognitive ability continuum. Therefore, IQs are not reported as continuous variables (scores such as 100.53 and 112.52).

In this section, we have seen how numbers are categorized into different scales of measurement: continuous variables, and discrete variables. Now we move on to the subject of data organization. We will see that teachers can

Figure 2.1

Continuous Variable

Time in Minutes
10:25.................................12:50.................................14:75

10...........11...........12...........13...........14...........15
Time in Hours

arrange test data so that they can be organized and interpreted descriptively through the design of statistical graphs.

ORGANIZING AND ARRANGING DATA

Frequency Distributions. After collecting test results, teachers need to organize and arrange test data into a systematic numerical arrangement. First, you should begin data organization by arranging the numbers in order, which is called an ***array***. For example, as a science teacher, you administer a test to 23 students ($n = 23$). The students' scores on the test are shown in Table 2.2.

The next step is to organize these test results so that they are more understandable. You do this by arranging the scores from the highest to the lowest (see Table 2.3). This pattern is the array. An array makes it simple for you to visualize the entire range, or distribution, of scores.

CHAPTER 2

Frequency Table. Once we have created an array, we can create a *frequency table*. A frequency expresses the frequency of occurrence. For example, a teacher can see how many students out of 20 obtained 75% of the possible grade on a social studies test. Frequency tables are useful when working with large distributions of numbers because they simplify data analysis by organizing the data. To create a frequency table from the array shown in Table 2.3, we tally the scores on the basis of each value's frequency (see Table 2.4, p. 14). Each check mark indicates one appearance of a particular score in the array.

Table 2.3 An Array for Science Test Results
100
98
97
90
89
89
89
81
80
80
79
77
77
74
73
72
72
70
67
65
64
50
42

Check for Understanding

2.1. Create an array for the following test scores ($n = 20$): 89, 85, 80, 99, 98, 64, 50, 42, 90, 90, 90, 80, 79, 77, 77, 74, 73, 72, 70, 80.

2.2. A questionnaire has items concerning (a) the number of elementary teachers in a school district, (b) the county in which the district is located, and (c) the district's rank on a recent statewide assessment tool. Identify the scale of measurement (ordinal, nominal, interval, or ratio) used for each of the above pieces of data.

2.3. Which scale would be used to determine the time in seconds that it takes for a student to complete a problem-solving exercise?

2.4. Define the term *ordinal scale* and give an example of an ordinal scale that teachers may use.

Frequency Graphs. *Frequency graphs* present frequency distributions, which are a useful way to organize data into graphs and tables. First, to help you understand graphs, we should review the basic algebraic principles that apply to the graphical presentation of data. You probably recall that when graphing or plotting is done with reference to two lines (which are referred to as *coordinate axes*) the horizontal axis is known as the *x*-axis and the vertical axis is called the *y*-axis. These basic lines are perpendicular to each other. Figure 2.2 represents a system of coordinate axes, and Figure 2.3 (p. 15) shows how these axes would be arranged on a *Cartesian grid* (named for the French philosopher Rene Descartes). You may be familiar with the Cartesian grid example found in Figure 2.3, which shows *x*- and *y*-axes that range from negative to positive values.

The first type of statistical graph that we will discuss is the frequency polygon, a type of line graph that shows the frequency of each data point. Figure 2.4 is an example of a frequency polygon.

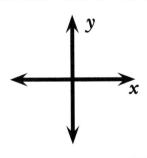

Figure 2.2 Coordinate Axes

Table 2.4

Creating a Frequency Table From the Array Shown in Table 2.3

(1) Group the Numbers	(2) Tally the Numbers	(3) Determine the Frequency
100	✓	1
98	✓	1
97	✓	1
90	✓	1
89, 89, 89	✓ ✓ ✓	3
81	✓	1
80, 80	✓ ✓	2
79	✓	1
77, 77	✓ ✓	2
74	✓	1
73	✓	1
72, 72	✓ ✓	2
70,	✓	1
67	✓	1
65	✓	1
64	✓	1
50	✓	1
42	✓	1

The second type of statistical graph that we will examine is the histogram, which is a type of bar graph. The width of each bar covers the numerical value of the scores. This type of graph is illustrated in Figure 2.5 (p. 16).

The next graph we look at is a simple bar graph, a statistical graph that is typically used when a frequency distribution is displaying data from nominal or ordinal data scales. This type of graph is illustrated in Figure 2.6 (p. 16). Note that the height of each bar corresponds to the frequency of the numerical value.

Check for Understanding

2.5. What type of graph would be appropriate for displaying the distribution of grades for 20 students in a mathematics class? Assume that the letter-grade distribution is the following for the students: As = 3, Bs = 5, Cs = 7, Ds = 3, Fs = 2.

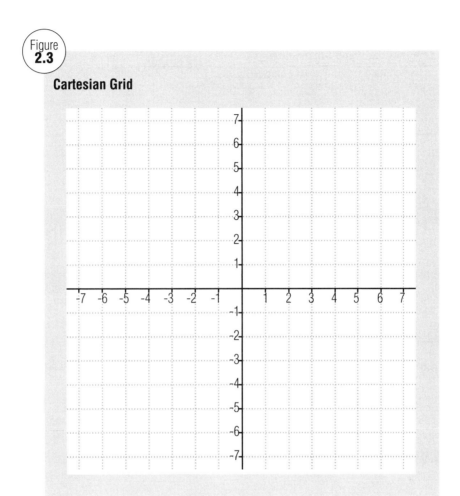

Figure 2.3 Cartesian Grid

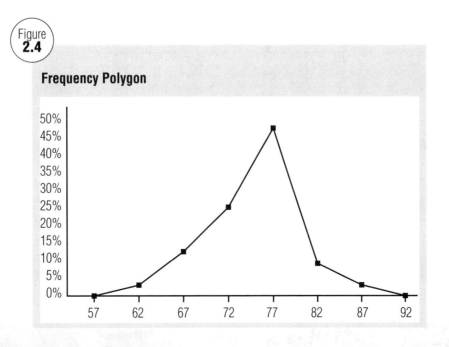

Figure 2.4 Frequency Polygon

2.6. Draw a bar graph for the distribution of the following SAT-I mathematics scores: 390, 600, 440, 700, 390, 600, 750, 600, 390, 500, 500, 500, 300, 700, 500, 440, 440, 500, 600, 780. Hint: You should organize the scores into a frequency table.

2.7. Explain why a teacher might decide to create a frequency graph to illustrate test results.

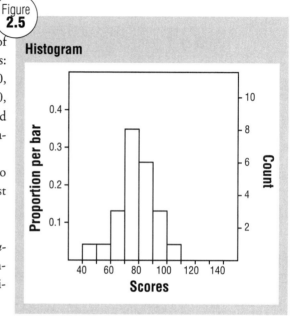

Figure 2.5

Normal Distributions. A *normal distribution* is a distribution in which scores are concentrated close to the center of a symmetrical distri-

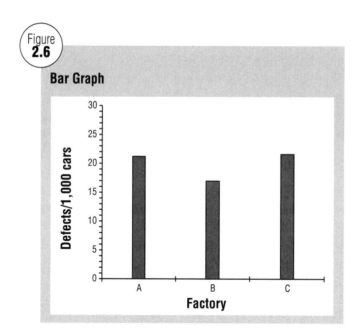

Figure 2.6

bution of scores, with only a few scores at either extreme. The normal distribution is often referred to as a **bell curve** because of its shape (see Figure 2.7).

All normal distribution graphs have the same general form in that they are symmetrical, and the scores concentrate closely around the center, taper off from the center high

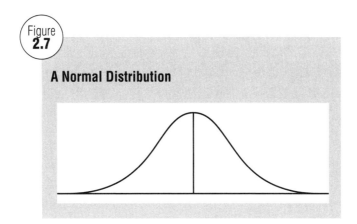

Figure 2.7

A Normal Distribution

point, and move toward the left and the right bases. Observe that the extreme left, which is known as the left tail, has relatively few scores; and, as we move to the right, toward what is known as the right tail, the number of scores increases progressively to a maximum number or frequency at the middle position. As we continue to move to the right of the middle position, the number of scores progressively decreases as we move nearer to the right tail.

An understanding of the characteristics of normal distribution is essential for a teacher because IQ (e.g., Stanford-Binet Scale) and standardized test scores (e.g., Iowa Tests of Basic Skills) are normally distributed. Again, a normal distribution occurs when many scores are clustered in the middle and fewer scores fall within the tails, the outside limits of the normal distribution. An analogy of height and intelligence is helpful. Very few people are three-feet tall or seven-feet tall, and very few people have either extremely high intelligence or extremely low intelligence. Most people are somewhere in the middle of each of these categories—about 68% of all people fall in the middle of the normal distribution. More time will be dedicated to the discussion of the bell curve, also known as the normal curve, in subsequent chapters.

Skewed Distributions. Sometimes a distribution will have more scores clustered at either the high or the low ends, which shifts the center of the distribution to one side or the other. When the scores are clustered at the high end of the scale, distributions are *negatively skewed.* The term *negatively* can be misleading because we generally associate *negative* with low and *positive* with high. In this instance, the associations are reversed. For example, if the majority of students receive As and Bs on a test, rather than Cs, Ds, or Fs, the distribution is negatively skewed. An example of a negatively skewed distribution is shown as the grade inflation of Figure 2.8 (p. 19).

The opposite is a positively skewed distribution, which occurs when the scores are crowded together at the low end (or left) of the scale. An example of a *positively skewed* distribution of scores is income, because more people have low incomes than high income. An example of a positively skewed distribution is shown in Figure 2.9 (p. 19).

CHAPTER 2

> ### The Bell Curve Controversy
>
> In 1994, two Harvard University researchers, Richard Herrnstein and Charles Murray, wrote ***The Bell Curve: Intelligence and Class Structure in American Life***. In this book, the authors found that intelligence level is related to social-class rank: People with high intelligence are more likely to occupy high social positions in occupations such as physician, lawyer, and professors. Herrnstein and Murray contend, for example, that the emergence of a "cognitive elite" class of individuals has transformed the social and economic boundaries of society and that people with high intelligence further their education and gravitate toward higher-paying jobs. In contrast, Herrnstein and Murray present a correlation between people with low intelligence and social problems such as school dropouts, high unemployment, domestic violence, out-of-wedlock births, work-related injury, and crime. In addition, the authors said that measured levels of intelligence differ among ethnic groups, and that Asian Americans, for example, score higher on IQ exams and have higher IQs than African Americans.
>
> In 1996, the late Steven Jay Gould, a well-known Harvard professor and science writer, responded to Herrnstein and Murray's book with the republication of his book ***The Mismeasure of Man: The Definitive Refutation to the Argument of the Bell Curve***, in which he explores the history of intelligence theory from the late 1700s to the present. Gould argued that race and class differences cannot be explained by categorical classifications and IQ test scores alone. Gould's conclusion was that Herrnstein and Murray were prejudiced against the lower social classes of society. As you can see from this well-known case, it is essential that you understand that statistics can be misleading, misunderstood, and abused. With this understanding, you should accept the necessity of accurately interpreting all statistical representations.

This concept will be explained further in the next chapter, when we explore central tendency. In this next chapter, you will also gain an understanding of how the position of each of the three measures of central tendency (the mean, the median, and the mode) changes from being at the same point in a normal distribution to different levels or points in a skewed distribution.

Check for Understanding

2.8. Explain a classroom scenario in which a negatively skewed distribution of grades would be likely.

2.9. Draw (1) a normal distribution, (2) a negatively skewed distribution, and (3) a positively skewed distribution.

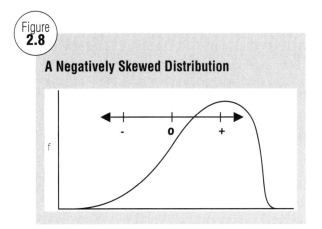

Figure 2.8 A Negatively Skewed Distribution

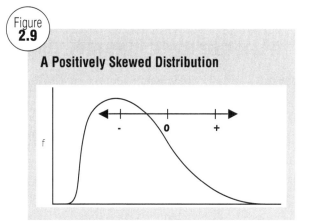

Figure 2.9 A Positively Skewed Distribution

2.10. If we had height measurements for every adult in the United States and plotted a curve with all of the data, would the distribution be normal or skewed?

Rounding Numbers

When performing statistical calculations, you determine the level of accuracy necessary for the problem. For example, scientists will sometimes carry their figures to four decimal places—88.6783. If rounded to two decimal places, this figure would become 88.68, and to one decimal, 88.7. For the most part, at least for the subsequent problems that you will encounter in this book, you and all educators should use a system for rounding numbers to two decimals. A suggestion for rounding to two decimals: When the third decimal is less than five, the second decimal does not change—74.983 becomes 74.98. When the third decimal is 5 or greater, the second decimal is increased by one—86.768 becomes 86.77 and 86.764 becomes 86.76.

SUMMARY

There are four different scales of measurement: ordinal, nominal, interval, and ratio. In educational statistics, we most often examine numerical data from the perspective of interval scales of measurement. An ordinal scale, for example, can show class rankings; a nominal scale is appropriate for dividing students according to gender; an interval scale can show the distribution of test scores; and a ratio scale can be used to illustrate the components of time. A continuous variable can be divided into an infinite number of fractional parts, whereas a discrete variable is not divisible. For example, temperature and money are continuous variables because they are divisible, while IQs are discrete variables because they are not divisible.

Descriptive statistics provide the basis for portraying data distributions through graphics such as histograms, bar graphs, and frequency polygons. A normal distribution of data is seen as symmetrical, but when the data are clustered at one end or the other, the distribution is skewed.

CHAPTER 2

CHAPTER REVIEW QUESTIONS

2.11. If a teacher arranges grades from highest to lowest, what scale of measurement is being used?

2.12. Organize the following test scores into a frequency array:
75, 100, 67, 88, 74, 91, 55, 100, 67, 74, 77, 82, 87, 94, 66, 71, 75, 84, 91, and 82.

2.13. Using the frequency array that you made in problem 2.12, create a frequency table.

The following table is for questions 2.14 and 2.15.

X	Frequency
97	1
86	2
84	1
75	3
74	1
68	1

2.14. For these data, n = ___.
   a. 5
   b. 6
   c. 9
   d. cannot be determined

2.15. For the set of data presented above, draw a histogram.

2.16. What is the shape of the distribution for the following set of test scores?
40, 50, 60, 60, 70, 70, 70, 80, 80, 90, 100
   a. positively skewed
   b. negatively skewed
   c. symmetrical
   d. cannot be determined

Questions 2.17, 2.18, and 2.19 refer to the following graph showing a distribution of test scores

2.17. This is an example of a _____ distribution:
   a. positively skewed
   b. negatively skewed
   c. symmetrical
   d. cannot be determined

2.18. Based on the individual scores represented in this graph, how many students took the test? (What is the value of n?)

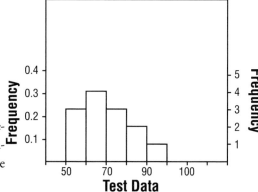

a. 13
b. 4
c. 5
d. cannot be determined

2.19. For this set of exam scores, how many students scored under 70?
a. 10
b. 5
c. 7
d. cannot be determined

The following graph is for question 2.20.

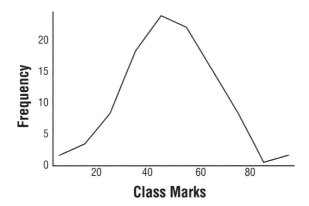

2.20. This graph is an example of a _____.
a. histogram
b. frequency polygon
c. bar graph
d. none of the above

ANSWERS: CHECK FOR UNDERSTANDING

2.1. 99, 98, 90, 90, 90, 89, 85, 80, 80, 80, 79, 77, 77, 74, 73, 72, 70, 64, 50, 42

2.2. a. ratio
b. nominal
c. ordinal

2.3. A continuous variable

2.4. An ordinal measurement assigns higher numbers to individuals who score higher on a test. The most common type of ordinal measurement is rank order. For example, a teacher might organize test scores from the highest score to the lowest. The higher the test score, the higher the rank.

2.5. A bar chart because the data scale is nominal.
2.6.

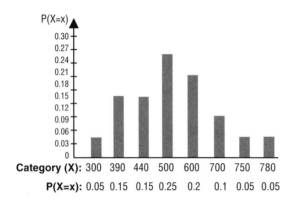

2.7. By organizing and simplifying your data into a graph, you condense them into a comprehensible snapshot.
2.8. A gifted or advanced placement class because the students are of above average aptitude and/or cognitive ability.
2. 9.

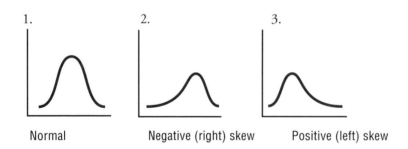

2.10. A normal distribution. Since we have an entire population, this distribution would be normal.

ANSWERS: CHAPTER REVIEW QUESTIONS

2.11. Ordinal
2.12. 100, 100, 94, 91, 91, 88, 87, 84, 82, 82, 77, 75, 75, 74, 74, 71, 67, 67, 66, 55

2.13.

X	Frequency
100	2
94	1
91	2
88	1
87	1
84	1
82	2
77	1
75	2
74	2
71	1
67	2
66	1
55	1

2.14. c. 9

2.15.

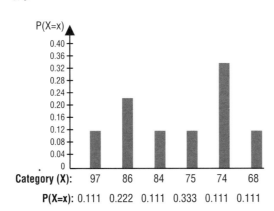

2.16. c. symmetrical
2.17. a. positively skewed
2.18. a. 13
2.19. c. 7
2.20. b. frequency polygon

CALCULATOR EXPLORATION

Using the TI-73, TI-83, or TI-84 graphing calculator, you can easily sort numbers by following these keystrokes.

TI-73

Step 1. Display list editor by pressing [LIST]. Under L1 enter Science Test Results in Table 2.3.

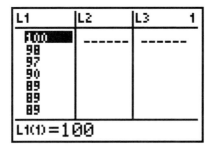

Step 2. Press 2nd [LIST] to activate [STAT]. Next, scroll to the right and select [OPS]. Next, scroll down to [1: Sort A (], press [1], then press [ENTER].

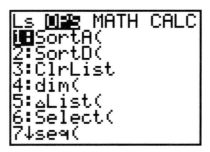

Step 3. After the [SortA(] is available, press 2nd [LIST] to activate [STAT]. Next, press [1], then press [)], and press [ENTER].

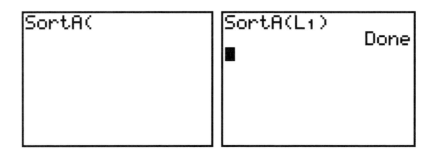

Step 4. Press [List] to get the variables listed from lowest to highest.

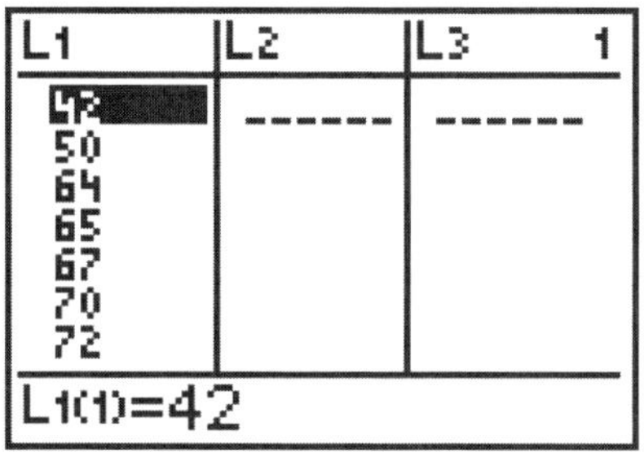

CHAPTER 2

[L1 shows: 42, 50, 64, 65, 67, 70, 72; L1(1)=42]

TI-83/TI-84

Step 1. First press [STAT]. Next, select [1:Edit] by pressing [1]. Here you can enter up to 999 elements.

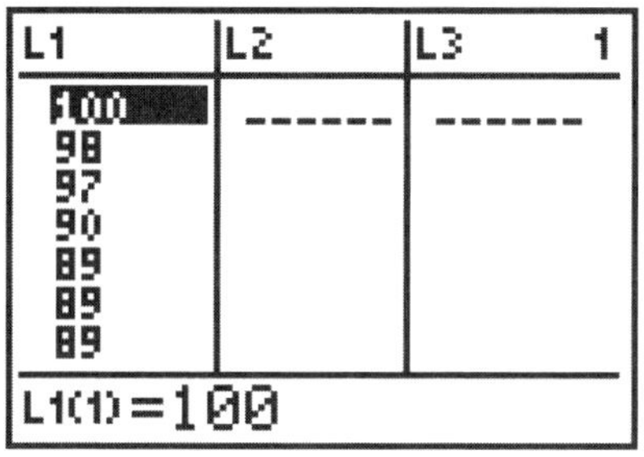

[L1 shows: 100, 98, 97, 90, 89, 89, 89; L1(1)=100]

Step 2. Press [STAT] and scroll down to [Sort A (]. Next, activate the 2nd function, press [L]1, and press [)], which is above the 9. Next press [ENTER].

[SortA(| SortA(L1) Done]

Step 3. Press [LIST]. Next, scroll down to [1: L1] and press [1]. Next, press [ENTER] to get the numbers listed in order from lowest to highest.

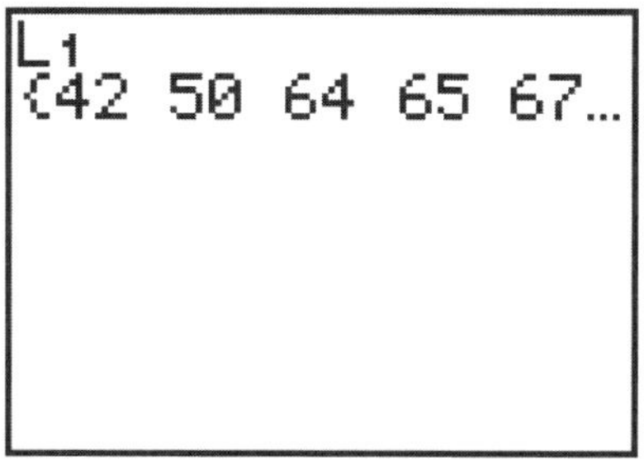

[L1 {42 50 64 65 67...]

INTERNET RESOURCES

This site of the National Center for Educational Statistics provides a kaleidoscope of statistics related to national and international performance in education. For example, How many schools have access to the internet? Which state scores the highest on the NAEP test? *http://nces.ed.gov*

The National Education Association website offers statistics related to education. Graphic displays include data on teacher salaries, public school enrollment, and expenditures on education. *ww.nea.org/edstats*

This site offers helpful suggestions for the understanding and appropriate implementation of elementary statistical calculations. *http://math.about.com/cs/statistics*

REFERENCES

Allen, M. J., and W. M. Yen. 2002. *Introduction to measurement theory.* Chicago: Waveland.

Gould, S. J. 1996. *The mismeasure of man: The definitive refutation to the argument of the bell curve.* New York: W. W. Norton.

Herrnstein, R. J., and C. Murray. 1994. *The bell curve: Intelligence and class structure in American life.* New York: Free Press.

FURTHER READING

Christmann, E. P. 2002. Graphing calculators. *Science Scope* 25 (5): 46–48.

Christmann, E. P., and J. L. Badgett. 2001. A comparative analysis of the academic performances of elementary education preprofessionals, as disclosed by four methods of assessment. *Mid-Western Educational Researcher* 14 (2): 32–36.

Christmann, E. P., and J. L. Badgett. 1999. The comparative effectiveness of various microcomputer-based software packages on statistics achievement. *Computers in the Schools* 16 (1): 209–220.

Elmore, P. B., and P. L. Woehlke. 1997. *Basic statistics.* New York: Longman.

Gravetter, F. J., and L. B. Wallnau. 2002. *Essentials of statistics for the behavioral sciences.* New York: West.

Raymondu, J. C. 1999. *Statistical analysis in the behavioral sciences.* New York: McGraw-Hill.

Thorne, B. M., and J. M. Giesen. 2000. *Statistics for the behavioral sciences.* Mountain View, CA: Mayfield.

Zawojewski, J. S., and J. M. Shaughnessy. 2000. Mean and median: Are they really so easy? *Mathematics Teaching in the Middle School* 57: 436–440.

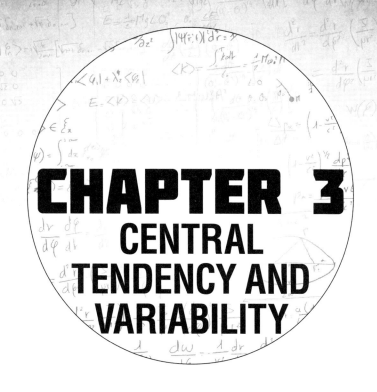

CHAPTER 3
CENTRAL TENDENCY AND VARIABILITY

OBJECTIVES

When you complete this chapter, you should be able to
1. formulate questions that can be addressed with data and collect, organize, and display relevant data to answer them;
2. use measures of central tendency, focusing on the mean, median, and mode;
3. describe the shape and important features of a set of data and compare related data sets, with an emphasis on how the data are distributed;
4. select and use appropriate statistical methods to calculate standard deviation; and
5. compare different representations of the same data and evaluate how well each representation shows important aspects of the data.

Key Terms

When you complete this chapter, you should understand the following terms:

central tendency	range
heterogeneous group	sample standard deviation
homogeneous group	skewed distribution
mean	standard deviation
median	variation
mode	variability
normal distribution	variance
outlier	

CHAPTER 3

This chapter presents two important interrelated topics in statistics: central tendency and variability. Measures of *central tendency* show how similar the data points in a set of data are, while measures of *variability* show how much the data points vary. Central tendency calculations determine representative scores within a distribution (see Chapter 2). For example, in a set of test scores consisting of 82, 86, 90, 94, and 98, the central score, 90, is the most representative of all the scores. The difference between the highest and lowest scores, 16, shows how much the scores vary. Variability shows how "scattered" or "spread out" the scores are around the center point of the distribution. In the first half of this chapter we will discuss three important measures of central tendency, the *mean*, *median*, and *mode*. In the second half, we will cover three measures of variation, the *range*, *standard deviation*, and *variance*.

Both central tendency and variation have important practical uses for teachers. For example, suppose you administered a 100-point science test to two classes of 25 students each (see Table 3.1). Based on the scores reported in Table 3.1, the average, or mean score for Class A is 77.12 ($n = 25$), and the mean score for Class B is also 77.12 ($n = 25$). Because the average scores are identical, you might assume that there is no difference between the two groups. If you examine variability as well as central tendency, however, you immediately see an important difference. For example, Class A's scores range from a low of 0 to a high of 100, while Class B's range from a low of 70 to a high of 84. Clearly, Class A's scores are more spread out, or have more variation.

Because Class A's scores have such high variation, the class is probably a group of mixed abilities. Groups of mixed ability are *heterogeneous*. In contrast, Class B's scores fall around the same point in the scale, reflecting a more similar range of ability. Groups that are of similar ability are *homogeneous*. The difference in variability would be important to you as a teacher, because you would need to use different instructional strategies with each group.

Table 3.1

Science Test Results	
Class A	Class B
100	84
99	83
98	82
98	81
97	80
94	80
93	79
88	79
87	78
86	78
85	78
80	77
79	77
78	77
77	77
75	76
74	76
70	76
69	75
67	75
66	74
65	73
54	72
49	71
0	70

CENTRAL TENDENCY

This section presents the three measures of *central tendency:* the *mean*, the *median*, and the *mode*. A measure of central tendency is a value that is representative of a data set. In a normal distribution of scores (see Chapter 2), all measures of central tendency would fall precisely at the same point, which is the line located exactly in the middle of the bell curve (see Figure 3.1, p. 30). Measures of central tendency are useful for summarizing a large set of scores in a meaningful way. We read and hear these terms all the time in the media:

- The *mean* is the average value of all the data in the set.
- The *median* is the value that has exactly half the data above it and half below it.
- The *mode* is the value that occurs most frequently in the set.

Some Common Measures of Central Tendency Found in the Media

- Average Teacher Salary in the United States: $45,930
  (NEA Research, Estimates databank, Fall 2003)

- U.S. median household income: $42,228
  (U.S. Census Bureau, 2002)

- Average household net worth of the top 1% of wage earners: $10,204,000
  Average net worth of the bottom 40% of wage earners: $1,900
  (Edward N. Wolff, "Recent Trends in Wealth Ownership, 1983–1998," April 2000)

- Definition of middle class in terms of mean annual income: $32,653 to $48,979
  (Economy.Com's The Dismal Scientist, 1999)

- Median hourly wage of a former welfare recipient: $6.61
  (Urban Institute, 2000)

- Bill Gates's average hourly wage: $650,000/hr
  (Bill Gates's Net Worth Page, average since 1986)

- Average teacher salaries
  U.S. average: $45,930
  California: $56,283 (highest)
  South Dakota: $32,416 (lowest)
  (NEA estimates for 2003)

CHAPTER 3

The Mean

The most common measure of central tendency is the *mean*, which is the arithmetic average of a set of scores. To compute the mean from a set of numbers, add them, then divide their sum, Σ_x, by the total number of values, n (see Table 3.2 for the definitions of Σ and other symbols). For example, if a student has obtained scores of 16, 18, and 17 on three math quizzes, the student's mean, or average score is 17. The symbol $\overline{x}$, which represents the sample mean, will be used in most cases to represent the mean in this text. The procedure is illustrated in Equation 3.1.

Figure 3.1

Bell Curve

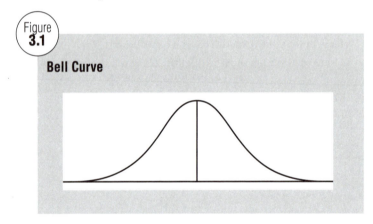

$$\overline{x} = \frac{\Sigma x}{n}$$

Equation 3.1

Table 3.2

Symbols Used to Compute Measures of Central Tendency

Symbol	Definition
Σ	The summation of a set of values
x	The variable that represents the individual data values
n	The number of values in a sample
N	The number of values in a population
$\overline{x}$	The mean of a set of sample values
Md	The median of a set of sample values
μ	The mean of all of the values in a population
s	The standard deviation of a set of sample values

National Science Teachers Association

CHAPTER 3

For example, you might administer a quiz to a small class of 10 students. If x represents an individual test result, and $x_1 = 96$, $x_2 = 96$, $x_3 = 87$, $x_4 = 84$, $x_5 = 79$, $x_6 = 75$, $x_7 = 72$, $x_8 = 68$, $x_9 = 65$, $x_{10} = 55$, then the sum of the $n = 10$ scores is $\Sigma x = 96 + 96 + 87 + 84 + 79 + 75 + 72 + 68 + 65 + 55 = 777$. Thus, the calculation of the mean, x, using equation 3.1, is $777/10 = 77.70$.

Because the mean takes into account the value of each score, one extremely low or high score can affect its value. For example, recall the student who scored 16, 18, and 17 on three math quizzes. If she scored a 2 on a fourth quiz, her mean score would decrease from 17 to 13.25. The mean is very sensitive to extreme measures, so it is not the best measure of central tendency to use in all cases.

Check for Understanding

3.1. Compute the mean for the following distribution of test scores: 100, 98, 97, 90, 89, 89, 85, 81, 80, 80, 79, 77, 77, 74, 73, 72, 72, 70, 67, 65, 64, 50, 42. 77

3.2. Add 5 points to each test score from 3.1 and compute the mean again. How did adding 5 points to each score affect the mean? 77 + 5 = 82

3.3. Add 10% to each score from 3.1. Explain what happened to the mean. Is there a difference between adding or subtracting from each score? Is there a difference between multiplying or dividing each score by a percentage?

The Median

The *median* of a set of scores is the value that divides the scores into equal halves. To return to the same simple example we used for the mean, if a student scores 16, 18, and 17 on three math quizzes, the student's median score is 17. To calculate a median, arrange the scores in rank order. The middle score is the median. For example, suppose a group of students obtains the following scores on a science quiz: 100, 94, 87, 85, 85, 79, 75, 71, 62. The median is the score that lies in the middle, 85 (Md = 85). For a distribution with an even number of scores (such as 95, 93, 88, 86, 77, 70), the median is halfway between the two middle values. It is calculated by averaging the two middle values (see Equation 3.2).

Equation 3.2

$$Md = \frac{88 + 86}{2} = 87$$

The median is most properly used when the distribution is skewed, or not arranged in a normal distribution (see Chapter 2). Unlike the mean, the median is not sensitive to an extremely high or extremely low score. Therefore, the median is a better way than the mean to describe this distribution. For example, assume that you wanted to compute the mean annual income of this particular class. A new student enrolls in the class. His name is Bill—Bill Gates. We find that the mean annual income of your class is now in the millions. But Bill Gates' income would not drastically change the median, so the median would give you a much more accurate idea of your class's average annual income.

Check for Understanding

3.4. Find the median for the following set of test scores: 100, 95, 93, 88, 86, 80, 77, 75, 70, 65.

3.5. Find the median for the following set of test scores (note: first arrange the scores in order from highest to lowest): 74, 100, 95, 86, 71, 74, 65, 55, 84.

3.6. Calculate the mean and median for the following test scores: 71, 100, 100, 98, 97, 99, 100, 88, 87, 10.

3.7. Which measure of central tendency better represents the distribution of scores, the mean or the median? Explain your answer.

The Mode

The *mode* is the score that occurs most frequently in a distribution. For example, if a student obtained scores of 97, 94, 88, 88, 80, 74, and 50 on spelling tests, the mode is 88, because this score occurs twice, which is more often than any of the student's other scores. A set of scores may have no mode, or it may have more than one mode. A distribution with two modes is called a *bimodal distribution*. For example, the following numbers represent quiz scores for an eighth-grade science class: 10, 10, 9, 9, 9, 8, 8, 8, 8, 7, 7, 7, 7, 6, 5, 3. This distribution is bimodal. The scores 7 and 8 both appear four times.

For this set of scores, the mean is 7.56 and the median is 8 (see Calculation 3.1). Thus, you can now see how all three measures of central tendency assist in the description of data. Because the median exceeds the mean in this set of scores, it is apparent that the distribution is negatively skewed (see Figure 3.2). As a teacher, for example, you should know that, contrary to what the term may imply, a negatively skewed distribution has more As and Bs than Ds and Fs. In contrast, a positively skewed distribution has more Ds and Fs than As and Bs. Remember, a negatively skewed distribution has more higher than lower scores, whereas a positively skewed distribution has more lower than higher scores. It is not unusual that a gifted class would have a negatively skewed distribution of test scores because the class comprises students with above average ability.

Calculation 3.1

$$Md = \frac{8+8}{2} = 8$$

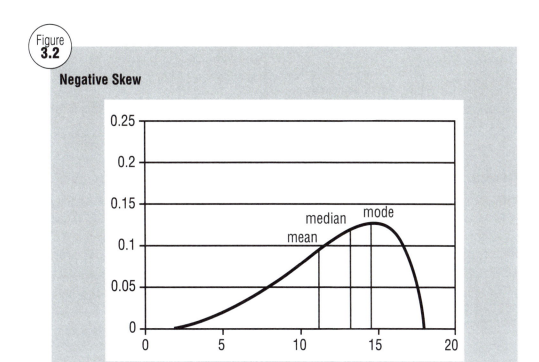

Figure 3.2

Negative Skew

Check For Understanding

Use the following distribution of test scores for questions 3.8 through 3.10: 95, 92, 92, 88, 85, 85, 85, 80, 77, 75, 75, 59.

3.8. What is the value of the mode?
3.9. What is the value of *n*? Calculate the mean and the median.
3.10. Describe the shape of the distribution.

Is Your Education Worth the Investment?

- Median household income for those with less than a 9th-grade education: $17,261
- Median household income for those with a 9th–12th grade education (no diploma): $21,737
- Median household income for high school graduates: $35,744
- Median household income for college graduates with a BA: $64,406
- Median household income for college graduates with an MA: $74,476
- Median household income for professional degree holders: $100,000
  (U.S. Census Bureau 1999)

VARIATION

Central tendency calculations help determine representative values of scores in a distribution. After calculating central tendency, the next step is to determine the *variation* of the distribution's scores. Variation provides a way to measure how scattered or spread out the scores are around the center point of the distribution. The three measures of variation we will discuss are the *range, standard deviation,* and *variance.*

Range

The *range* is the difference between the highest and the lowest scores in a data set. For an example, refer back to the science test scores for two classes given in Table 3.1. To calculate the range of scores in Class A, we subtract 0 from 100 and get a range of 100 (see Calculation 3.2). Likewise, for Class B we subtract 70 from 84 and arrive at a range of 14 (see Calculation 3.3). As you see, the range is a simple way to describe how spread out the scores are. Because the range does not take into account all of the scores in the distribu-

Calculation 3.2

Range = 100 − 0 = 100

Calculation 3.3

Range = 84 − 70 = 14

tion and is sensitive to extremely high and extremely low scores, it is not the most reliable measurement of variation, as you also can see from Calculation 3.2 and Calculation 3.3. Consequently, we use the variance and the standard deviation to provide more reliable and truthful measurements of variability.

Standard Deviation and Variance

The *standard deviation* is the best statistic to use to determine the mathematical variation, or the amount of dispersion, among the scores in relation to the mean of a distribution. It shows how spread out the data are in a sample or a population. The standard deviation is the first step in the calculation of most other statistics. In this text, we will cover only the *sample standard deviation,* which is what is used for the small samples that classroom teachers typically work with.

When teachers work with standardized tests, however, they use norm-referenced data. When we speak of *norm-referenced data,* we are talking about comparisons among the units within a set of data, which could be a set of test scores. With this approach, an individual's test score is viewed in terms of how it compares with the other test scores. For example, if an individual scored in the 95th percentile, we mean that he or she scored as high as or higher than 95% of the people who took the same test. Even if this student answered only 38 of 100 items correctly, he or she could still score in the 95th percentile, depending on how the other students scored. This student's raw score of 38, however, is viewed as *criterion-referenced data,* which is the number or percentage of items answered correctly out of 100 test items.

A simple way of viewing these two types of data illustrations is to understand that with criterion-referenced data we are comparing scores, whereas with norm-referenced data we are comparing the performances of people.

With norm-referenced data, the standard deviation is a point on a normal curve. Figure 3.3 (p. 36) shows the standard deviation units within a normal curve. The two areas in the middle area represent one standard deviation, the two areas lying directly beside the two middle areas represent two standard deviations, and the two outermost areas represent three standard deviation units. This concept will be useful in our discussion of standard scores in the next chapter.

Equation 3.4 (p. 37) shows the formula for calculating the sample standard deviation. In Table 3.2 (p. 30) we defined the symbol Σ, the uppercase Greek letter sigma, which means summation. (The lower-case sigma, σ, will be used as the symbol for population standard deviation. In this text, the symbol for sample standard deviation that will be used is the upper-case S.) Notice in Equation 3.5 (p. 38) that the sample standard deviation formula uses $n-1$ as the denominator in the formula. If you are wondering, $n-1$ is used as a mathematical method to correct for bias. It means that, if we have a small sample size, we will divide by a smaller number, which makes the product of the equation larger and thus allows the sample to reflect a less-biased estimate of the actual population.

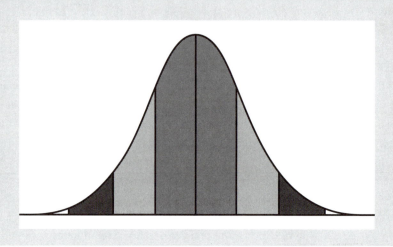

Figure 3.3

Standard Deviation and Normal Curve

Let's calculate the sample standard deviation of the scores for Class A in Table 3.1 (p. 28). First, we need to calculate the mean of this set of scores. Second, we calculate the deviation of each score from the mean: We subtract the mean from each score's value ($x -$ mean). Notice, if we were to add all of the deviations, the sum would equal zero, because some scores would be larger and some smaller than the mean. Therefore, when the deviations are summed, they equal zero (see the bell curve in Figure 3.1, p. 30). The third step is to square each deviation. The fourth step is to sum all of the squared scores. Calculation 3.4 shows the calculation of the standard deviation for the scores of Class A shown in Table 3.1. Then, the fifth step is to divide the sum of the squared score (i.e., $\Sigma (X - \text{mean})^2$) by $n - 1$.

The resulting value is the average of the squares of the deviation, also known as the *variance* (see Equation 3.3). Like standard deviation, the variance is used to explain how scores are different from one another. The variance, however, is a squared value. Therefore, the sixth and final step is to calculate the sample standard deviation, which is easily calculated by taking the square root of the variance (see Equation 3.4).

Equation 3.3

$S^2 =$ Variance

$$S^2 = \frac{\Sigma (X - \text{mean})^2}{n - 1}$$

Or

$$S^2 = \frac{10952.626}{25 - 1} = 456.359$$

Calculation 3.4

Class A Scores X	Deviation (X − mean)	Squared Deviation (X − mean)^2
100.000	22.880	523.494
99.000	21.880	478.734
98.000	20.880	435.974
98.000	20.880	435.974
97.000	19.880	395.214
94.000	16.880	284.934
93.000	15.880	252.174
88.000	10.880	118.374
87.000	9.880	97.614
86.000	8.880	78.854
85.000	7.880	62.094
80.000	2.880	8.294
79.000	1.880	3.534
78.000	0.880	0.774
77.000	-0.120	0.014
75.000	-2.120	4.494
74.000	-3.120	9.734
70.000	-7.120	50.694
69.000	-8.120	65.934
67.000	-10.120	102.414
66.000	-11.120	123.650
65.000	-12.120	146.894
54.000	-23.120	534.534
49.000	-28.120	790.734
.000	-77.120	5947.494
Mean = 77.120	Σ(X − mean) = 0	Σ(X − mean)^2 = 10952.626

Equation 3.4

S = Standard Deviation = $\sqrt{\text{variance}}$

Or $S = \sqrt{456.359} = 21.363$

Another method used to calculate sample standard deviation is the computational formula. The computational formula is easier to calculate with a basic handheld calculator (see Equation 3.5). The following is a six-step guide to calculating sample standard deviation with the computational formula (see Table 3.3).

Equation 3.5

$$S = \sqrt{\frac{\Sigma X^2 - (\Sigma X)^2/n}{n-1}}$$

Where

ΣX^2 = The sum of the squared scores.

$(\Sigma X)^2$ = The square of the sum of all of the scores.

n = The total number of scores used in the computation.

Table 3.3

Computational Formula for Standard Deviation of the Science Test Results for Class A

Step 1. Add all of the scores.
100.000 + 99.000 + 98.000 ... + 0.000 = 1928.000

Step 2. Square each score and add all of the squared values.
$100.000^2 + 99.000^2 + 98.000^2 + ... 0.000^2 = 159640.000$

Step 3. Square the sum from Step 1 and divide by the total number of students in the class.
$1928.000^2/25 = 3717184/25 = 148687.360$

Step 4. Subtract the value obtained in Step 3 from the value in Step 2.
159640.000 − 148687.360 = 10952.640

Step 5. Divide the value calculated in Step 4 by n − 1, which gives us the variance.
$S^2 = 10952.640/24 = 456.360$

Step 6. Take the square root of the variance that was calculated in Step 5. This is the sample standard deviation.

$S = \sqrt{456.359} = 21.363$

National Science Teachers Association

Used together, the mean and the standard deviation can give you a good idea of how students performed on a test. For example, if you know that your students have a mean score of 70 on a mathematics test and the standard deviation is 10, you can estimate that about 68% of the students scored between 60 and 80 on the test and about 95% of the students earned scores between 50 and 90 on this exam. (2-SD)

This is because the normal distribution is based on the standard deviation in the following way: In any normal distribution of test scores, approximately 68% of the test scores fall within one standard deviation of the mean, 95% of the test scores fall within two standard deviations of the mean, and 99.9% of the test scores fall within three standard deviations of the mean (see Figure 3.4, p. 40). In the class whose scores we have been examining, however, we have a relatively high standard deviation of 21.363. This is because of the single score of zero on the test. A single score that is located far from the rest of the data is called an *outlier*. As a general rule, outliers should be investigated. If, for example, the student who scored a zero on the test received the zero because he was absent on the day of the test, this score should not be included in the data set. Subsequently, for the remaining 24 scores the removal of the zero changes our standard deviation to from 21.363 to 14.382 and the mean from 77.120 to a 79.522. Now our test results are a better reflection of those students who took the test with approximately 68% of the 24 scores within one standard deviation of the mean. Note, however, that if a student took the test and scored a zero, that score should be included among the results.

The following example will further illustrate the usefulness of the standard deviation. Assume that we have the following test scores from two sets of students:

Group A	Group B
50	50
50	40
50	60
50	70
50	30

The mean scores of the two groups is the same, but their standard deviations are quite different.

Check For Understanding

3.11. What is the relationship between the standard deviation and variance?
3.12. Calculate the range, variance, and standard deviation for Class B's science test results from Table 3.1 (p. 28).
3.13. For the sample of scores 5, 2, 5, 4, calculate the range, variance, and standard deviation.

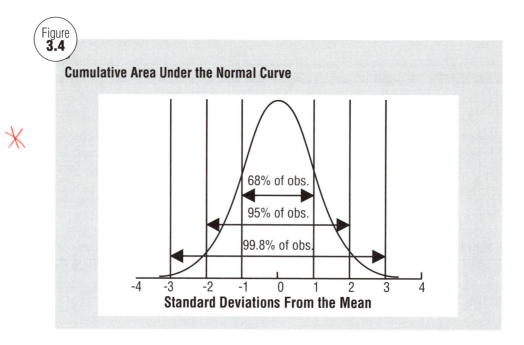

Figure 3.4

Cumulative Area Under the Normal Curve

SUMMARY

There are three standard measures of central tendency. In most situations, the mean, or arithmetic average, is the most commonly used measure of central tendency. The median is the exact middle point of a distribution. In a perfect normal distribution, shown in Figure 3.4, the mean, median, and mode all have the same value. In a skewed distribution, however, the mean value does not correspond with the other two measures of central tendency. As the "balance point," the mean is pulled toward the drawn-out tail of the skewed distribution.

The mode is the most frequently occurring score, and it is the least useful measure of central tendency when the mean or median can be used. The mode, however, tells us the score with the highest frequency.

The range, the standard deviation, and its square, the variance, are used to illustrate the spread, or variability, of a group of scores. The range is the measure of variability that shows the spread of the scores and informs a teacher of whether or not the test results include extreme scores.

CHAPTER REVIEW QUESTIONS

3.14. What is the value of the mean for the following set of test scores?
Test Scores: 100, 97, 93, 88, 84, 79, 75, 72, 68, 62, 50, and 0.
a. 85.50
b. 83.00
c. 71.44
d. 72.33

3.15. What is the value of the median for the following set of test scores?
Test Scores: 100, 97, 93, 88, 84, 79, 75, 72, 68, 62, 50, and 0.
a. 75.00
b. 79.00
c. 78.50
d. 77.00

3.16. What is the value of the mode for the following set of test scores?
Test Scores: 100, 97, 93, 88, 84, 79, 75, 72, 68, 62, 50, and 0.
a. 84.00
b. 79.00
c. 79.00 and 75.00
d. There is not a mode in this distribution of test grades.

3.17. Define and compare the three measures of central tendency.

3.18. Mrs. Smith's science class obtained the following scores on the Stanford Achievement Test: 100, 98, 97, 89, 89, 85, 82, 78, 77, 75, 75, 75, 68, 65, 60, 59, 30.
Determine the following:
a. $n =$
b. $S x =$
c. $x =$
d. median =
e. mode =

Match the characteristics listed in Column A with the terms in Column B. The options in Column B may be used more than once.

Column A

_____ 3.19. the score that half of the examinees score at or below

_____ 3.20. the arithmetic average of scores

_____ 3.21. the most frequently occurring score

_____ 3.22. also known as the 50th percentile.

_____ 3.23. more than one possible in the same distribution

Column B

a. mean
b. median
c. mode

3.24. From this set of test scores —100, 100, 98, 92, 90, 88, 84, 84, 84, 84, 76, 72, 66, 60, 55, 0— compute the following measurements:
   a. n =
   b. mean =
   c. median =
   d. mode =
   e. range =
   f. variance =
   g. standard deviation =

3.25. What statistical term do we use to describe how spread out or dispersed scores are within a distribution?

3.26. If a distribution of test scores has a small standard deviation with a compressed appearance, it is known as which type of group?
   a. Symmetrical group
   b. Homogeneous group
   c. Heterogeneous group
   d. Similar group

3.27. Define and compare the three major indicators of variability.

3.28. Which set of test scores has the highest variability?
   a. 77, 79, 82, 88
   b. 65, 75, 85, 95
   c. 70, 80, 90, 100
   d. 0, 40, 60, 100

3.29. To compute the sample variance, the sum of the squares is divided by:
   a. $n - 1$.
   b. N.
   c. n.
   d. $n - 1$.

3.30. The sample standard deviation is identified by which of the following symbols?
   a. S^2
   b. x
   c. S
   d. N

3.31. Determine the range from the following set of test scores: 87, 76, 43, 97, 99.

3.32. Calculate the standard deviation and variance of the following set of test scores: 100, 99, 90, 99, 100.

3.33. Based on the calculated standard deviation from question 3.32, is the group of scores heterogeneous or homogeneous?

Challenge Questions

3.34. If the average (arithmetic mean) of five numbers is equal to the median of the numbers, what is one possible value of *x* if the five numbers are 10, 6, 10, 4, and *x*?

3.35. Using the data from Class A's science test scores from Table 3.1 (p. 28), add 5 points to every score and then compute the mean, range, and standard deviation. How is the standard deviation affected when a constant is added to every score?

ANSWERS: CHECK FOR UNDERSTANDING

3.1. Mean = 77.00, sum X = 1771.00, n = 23

3.2. Mean = 82.00, sum X = 1886.00, n = 23
If a constant is added to a score, the same constant will be added to the mean.

3.3. Mean = 84.70, sum X = 1948.10, n = 23
Multiplying by a constant value is a method for changing the scale of measurement. When every score is multiplied by a constant value, the mean changes along with the constant value.

3.4. Median = $\dfrac{86 + 80}{2}$ = 83

3.5. 100, 95, 86, 84, 74, 74, 71, 65, 55
Median = 74

3.6. 100, 100, 100, 99, 98, 97, 88, 87, 71, 10
Mean = 85.0
Median = 97.50

3.7. Since the distribution is negatively skewed, the median of these scores would be the measure of central tendency that best depicts the distribution.

3.8. Mode = 85

3.9. n = 12, mean = 82.33, median = 85.00

3.10. Because the median exceeds the mean, the distribution is negatively skewed.

3.11. Standard deviation is the square root of variance.

3.12. Range is 14.000.
Variance is 12.524.
Standard deviation is 3.339.

3.13. n = 4
Range is 3.000.
Variance is 2.000.
Standard deviation is 1.414.

ANSWERS: CHAPTER REVIEW QUESTIONS

3.14. d. 72.33

3.15. d. 77.00

3.16. d. There is not a mode in this distribution of test grades.

3.17. The mean is the average value of all the data in the set. The median is the value that has exactly half the data above it and half below it. The mode is the value that occurs most frequently in the set.

3.18. a. $n = 17$
b. $S\,x = 1302$
c. $x = 76.59$
d. median = 77.00
e. mode = 75.00

3.19. b

3.20. a

3.21. c

3.22. b

3.23. c

3.24. a. $n = 16$
b. mean = 77.06
c. median = 84.00
d. mode = 84.00
e. range = 100
f. variance = 603.68
g. standard deviation = 24.57

3.25. Variation

3.26. b. Homogeneous group

3.27. The range is the difference between the highest and the lowest scores in a data set. The standard deviation is the best statistic to use to determine the mathematical variation, or the amount of dispersion, among the scores in relation to the mean of a distribution. The variance is the resulting value of the average of the squares of the deviation.

3.28. d. 0, 40, 60, 100

3.29. a. $n - 1$.

3.30. c. S

3.31. Range = 56

3.32. Sd = 39.11 and Variance = 1530.30

3.33. More homogeneous

ANSWERS: CHALLENGE QUESTIONS

3.34. $x = 0$ or $x = 20$ or $x = 7.5$

3.35. $n = 25$
Range = 100.000
Mean = 82.120
Standard deviation = 21.363.
Note: Although the mean changes, adding a numerical value to each score does not change the average distance of each score from the mean. Therefore, the variance is not affected.

CALCULATOR EXPLORATION

Using the TI-73, TI-83, or TI-84 graphing calculator, you can easily calculate the mean and standard deviation by following these keystrokes.

TI-73 Calculator

Step 1. Display list editor by pressing [LIST]. Under [L1] enter all 25 scores from class A in Table 3.1.

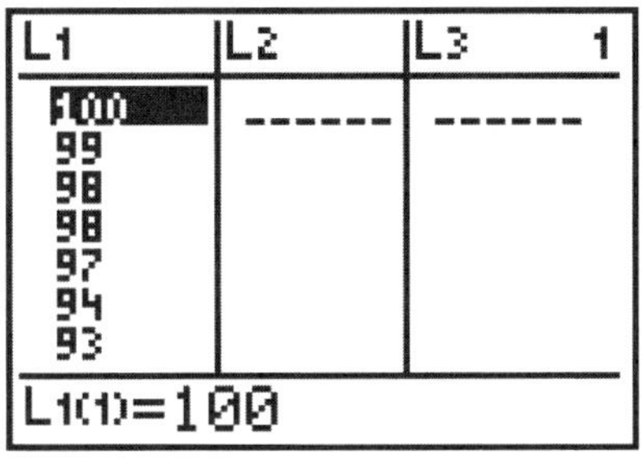

Step 2. Press 2nd [LIST] to activate [STAT]. Next, scroll to the right and select [CALC]. Next, scroll down to [1: 1 – Var Stats], press [1], then press [ENTER].

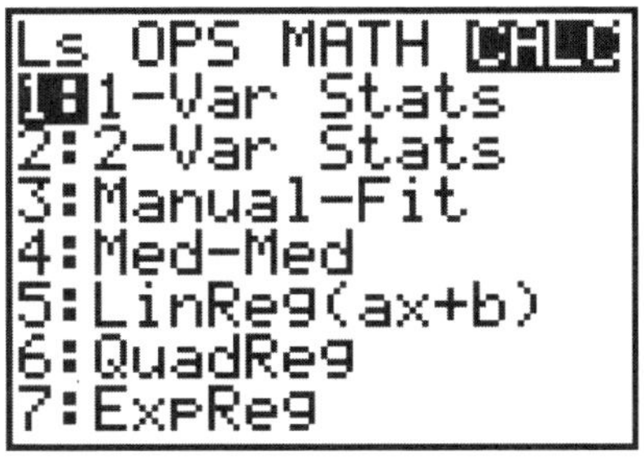

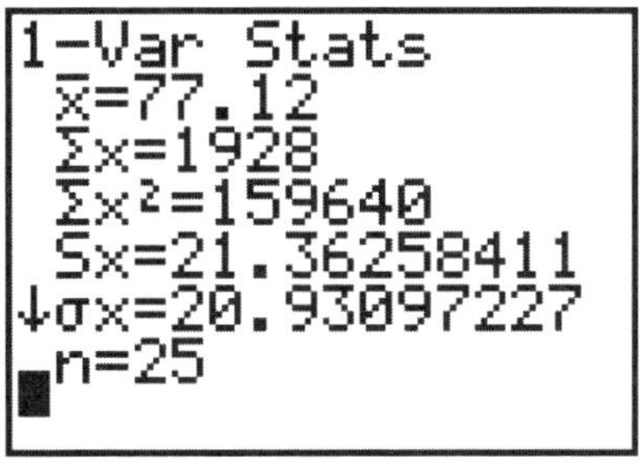

TI-83/TI-84 Calculator

Step 1. First press [STAT]. Next, select [1: EDIT] by pressing [1]. Here you can enter up to 999 elements. Under [L1] enter all 25 scores from class A in Table 3.1.

Step 2. Press [STAT] and scroll to [CALC]. Press [1: 1 – Var Stats].

Step 3. Press [ENTER] to get your results.

INTERNET RESOURCES

This section of the St. John's University website is a tool to calculate mean and standard deviation. After entering between 3 and 99 values, this program calculates mean and standard deviation. It also has an option that can import data from your computer, using copy and paste, and that can handle up to 1,024 numerical values. *www.physics.csbsju.edu/stats/cstats_NROW_form.html*

This section of *edhelper.com* is a tutorial on mean and standard deviation. At the end of the tutorial, the website creates a practice worksheet and e-mails you the answers to the problems. This is an excellent resource for practicing your calculations. *www.edhelper.com/statistics.htm*

This section of the website of Statistics Canada instructs users on the concept of standard deviation and variation. Also, this website provides links to the Canadian Census and other descriptive statistics that are compiled in Canada. *www.statcan.ca/english/edu/power/ch12/variance.htm*

FURTHER READING

Allen, M. J. 2002. *Introduction to measurement theory.* Prospect Heights, IL: Waveland.

Christmann, E. P. 2002. Graphing calculators. *Science Scope* 2 (5): 46–48.

Christmann, E. P., and J. L. Badgett. 2001. A comparative analysis of the academic performances of elementary education preprofessionals, as disclosed by four methods of assessment. *Mid-Western Educational Researcher* 1 (2): 32–36.

Christmann, E. P., and J. L. Badgett. 1999. The comparative effectiveness of various microcomputer-based software packages on statistics achievement. *Computers in the Schools* 16 (1): 209–220.

Elmore, P. B., and P. L. Woehlke. 1997. *Basic statistics.* New York: Longman.

Gravetter, F. J., and L. B. Wallnau. 2002. *Essentials of statistics for the behavioral sciences.* New York: West.

Raymondu, J. C. 1999. *Statistical analysis in the behavioral sciences.* New York: McGraw Hill.

Thorne, B. M., and J. M. Giesen. 2000. *Statistics for the behavioral sciences.* Mountain View, CA: Mayfield Publishing.

Zawojewski, J. S., and J. M. Shaughnessy. 2000. Mean and median: Are they really so easy? *Mathematics Teaching in the Middle School* 5 (7): 436–440.

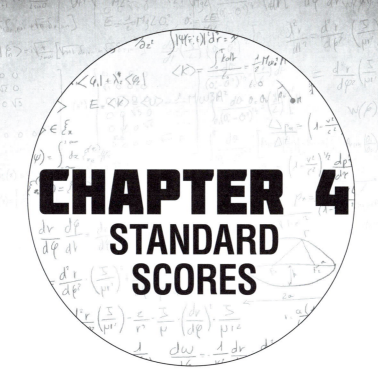

CHAPTER 4
STANDARD SCORES

OBJECTIVES

When you complete this chapter, you should be able to
1. demonstrate an understanding of percentile ranks and their relevance to classroom teachers;
2. compare the relationship between percentile ranks, standard scores, and the normal curve;
3. calculate percentile ranks from classroom and standardized test results;
4. find, use, and interpret measures of standard scores; and
5. compare and contrast T-scores, raw scores, and recentered test scores.

Key Terms

When you complete this chapter, you should understand the following terms:

centered score	standard score
percentile rank	T-score
raw score	z-score
relative standing	

Standard scores show an individual's relative performance within a group. We are all familiar with standard scores and use them all the time: Your scores on the SAT are standard scores, as are individual scores on achievement tests, class rank, intelligence tests, and a variety of other aptitude tests. Standard scores will help you understand your students' standardized test results and explain the results to parents. Granted, you will view most of the results of your own classroom assessments as criterion-referenced data (number or percentage of items correct).

CHAPTER 4

You must understand norm-referenced data (comparisons among local, state, and national scores), however, so that you can understand and then explain your students' and school's standardized test results. For example, if your school moves from the 10th to the 45th percentile as part of the Adequate Yearly Progress (AYP) mandated by the No Child Left Behind Act, you will be able to explain your school's position in terms of its progress rather than having to rationalize why your school is below the 50th percentile. In addition, you would be able to justify your school's "inadequate" AYP if it ranks in the 95th percentile for two successive years.

In this chapter we will discuss the following types of standard scores: ***percentile ranks***, ***z-scores***, and ***T-scores***. All are based on concepts—such as the mean, the normal distribution, and the standard deviation—already familiar to you from the last two chapters. In the case of class rank, however, keep in mind that a percentile rank from an ordinal scaled score or a rank score should be used. Z-scores, which are standard scores representing the number of standard deviation units a raw score is above or below the mean and which are used with normative data, are converted to percentile ranks when an individual score can be compared to a normal population (e.g., z-scores are calculated when the mean and the standard deviation from a population are known).

> ### Standard Scores in Schools
>
> Results on standardized tests are given as standard scores. On the SAT, the scores range from 200 to 800 with percentile ranks ranging from 1 to 99. Higher scores result from correctly answering more test questions. Both the score and the percentile rank compare the test taker's results with those of a recent representative national sample of high school students. A student's percentile rank shows the percentage of the test takers who earned scores at or below those of the student. For example, if a candidate's percentile rank is 70, the candidate's score is equal to that earned by 7 of 10 SAT test takers.

PERCENTILE RANKS

Percentile rank represents a student's position in a group relative to the number of students who scored at or below the position of that student. School districts use percentile ranks to calculate a student's rank within a class and to determine a student's relative standing on a standardized achievement or aptitude test. Test companies often report students' achievement test scores as percentile ranks. If a student scored at the 65th percentile on the reading test of the Stanford achievement tests, that student obtained scores as high as or higher than 65% of the students who took the test. To calculate percentile ranks, we use the concepts of the mean, standard deviation, and the normal distribution, as discussed in other chapters.

CHAPTER 4

Class ranks are based on students' grade point averages, or GPAs. The first step in calculating percentile ranks, based on GPAs or any other score, is to arrange the scores in order from highest to lowest, as shown in Table 4.1. Data organized in this way, from highest to lowest or lowest to highest, are referred to as *ordinal data*.

Table 4.1

Grade Point Averages (GPAs) for a High School Senior Class

3.9798	3.537	2.7129	1.6813
3.9321	3.5203	2.6584	1.573
3.9147	3.5177	2.6346	1.54
3.8922	3.4266	2.5612	1.5259
3.881	3.4249	2.5272	1.5
3.8761	3.419	2.4242	1.4375
3.8755	3.4072	2.3673	1.4341
3.875	3.3314	2.3386	1.1324
3.8593	3.1734	2.3253	1.1226
3.8511	3.1492	2.2799	1.063
3.8457	3.1296	2.2409	0.9971
3.8358	3.0909	2.1742	0.9854
3.8295	3.0849	2.1275	0.9752
3.7991	3.0817	2.0535	0.8887
3.7926	3.0787	2.002	0.7889
3.7558	2.9518	1.9785	0.7771
3.6858	2.9485	1.9559	0.7456
3.6827	2.9479	1.949	0.6664
3.6752	2.9294	1.899	0.6558
3.6685	2.9229	1.8635	0.5555
3.646	2.8504	1.8366	0.4889
3.6305	2.8444	1.7358	0.488
3.63	2.7992	1.7347	0.4526
3.61	2.7481	1.7154	0.225
3.6023	2.7264	1.7008	0

CHAPTER 4

Once the data are organized in rank order (ordinally), a percentile rank (PR) can be calculated. Equation 4.1 is the formula used to calculate percentile rank (PR) from ordinal data. Percentile ranks are reported on a continuous 100-point scale.

$$PR = 100 - \frac{(100R - 50)}{N}$$

Equation 4.1

In Equation 4.1, "R" is the rank position and "N" is the sample size. Therefore, if we use Equation 4.1 to analyze the GPA data from Table 4.1, the student who ranked first in a class of 100 would be at the 99.5th percentile rank (see Calculation 4.1)

$$PR = 100 - \frac{(100 \times 1) - 50}{100} = 99.5 \text{th PR}$$

Calculation 4.1

Likewise, the student in the 50th rank position would be at the 50.5th percentile rank (see Calculation 4.2).

$$PR = 100 - \frac{(100 \times 50) - 50}{100} = 50.5 \text{th PR}$$

Calculation 4.2

Notice that the equation estimates the percentile rank as based at the midpoint of the interval, i.e., the midpoint of the interval (50 – 51) is 50.5. That means the percentile rank of the student in the 100th rank position would be the 0.5th percentile rank (see Calculation 4.3).

$$PR = 100 - \frac{(100 \times 100) - 50}{100} = 0.5 \text{th PR}$$

Calculation 4.3

Equation 4.1 is very useful for a guidance counselor who is interested in calculating the class rank of students within a particular senior class when a normal distribution is not available. Moreover, school districts sometimes report progress on the basis of percentile ranks for scholarship eligibility and college admissions decisions.

Francis Galton (1822–1911)

Francis Galton, Charles Darwin's cousin, became interested in measuring differences in the cognitive, affective, and psychomotor characteristics of people in England in the late 1800s. One of his studies reported measurements of the stature of 8,585 adult men. As reported by Galton, the mean height for men was 67.02 in., with a standard deviation of 2.564 in. Galton, being a statistician, plotted his findings as a frequency polygon, which showed that the results formed the shape of a normal curve (see Figure 4.1, p. 54). As a result of Galton's research, scientists measure all sorts of other characteristics, such as weights, skull sizes, and reaction times.

Concurrent with Galton's research, an increasing number of students were attending common schools throughout the United States from the mid to late 1800s. Increased student attendance at common schools, which are known today as public schools, created a need for more teachers, resulting in the first teacher preparation institutions, called normal schools, opening their doors throughout the country.

We see that the similarities between the term normal school and the statistical term normal distribution are no coincidence when we consider that they evolved almost simultaneously with the norm-referenced testing theories of G. Stanley Hall, Charles Spearman, and E. L. Thorndike. As discussed in Chapter 2, normal distribution, which is also known as the normal curve or bell curve, refers to a distribution that is symmetrical: The areas on both sides of the curve are identical.

NORMAL DISTRIBUTIONS AND PERCENTILES

To visualize a normal distribution, imagine being able to categorize each man on Earth along a parallel line according to height. Theoretically, if we were able to create stacks of their bodies according to their heights, we would see a mountain of men in the shape of a normal distribution (see Figure 4.1, p. 54, for how this distribution would look). Keep in mind that, according to the Guinness Book of World Records (2002), the tallest living man is 7 ft. 8.9 in., and the shortest man is 2 ft. 4 in. tall.

Recalling our discussion on central tendency, you will remember that when we are moving from left to right, the distribution is formed by a gradual movement upward from the extreme short height of 2 ft. 4 in. to the peak average height of about 5 ft. 7 in. Then the curve would gradually slope downward as the frequency of taller men diminished from the extreme height of 7 ft. 8.9 in. If we use Galton's statistics, the standard deviation for male height is about 2.56 in., with a mean height of 5 ft. 7 in., which is about the same as what we find in the United States today. Therefore, men who range in height between 5 ft. 4 in. and 5 ft. 10 in. are not considered unusual.

Applied to industry, most manufacturers of furniture, automobiles, beds, and other goods build consumer products within a range of 2 or 3 full standard deviations of human sizes, such as height, weight, and arm length. This is because, mathematically, three full standard deviations below and above the mean on a normal distribution are equivalent to about 99% of the normal curve area; or, in this case, 99% of the male population. This is mass production, or standardized production for a mass market.

More practical for the classroom teacher is being able to use percentile ranks and the normal distribution together to analyze a student's scores on different tests. The ability to interpret scores on standardized tests, such as the Wechsler Intelligence Scale for Children III (WISC III) and the SAT is essential. Moreover, being able to determine the percentile rank of an individual student from a *raw score*, the unadjusted score on a test, is a skill every teacher should have, because comparisons between IQ scores and standardized test results on the same scales will help you make relative comparisons between and among a variety of aptitude and achievement tests. For example, you can interpret the relative classroom performance of a youngster on the basis of a comparison among his or her results on an achievement test, grades, and, possibly, IQ test scores.

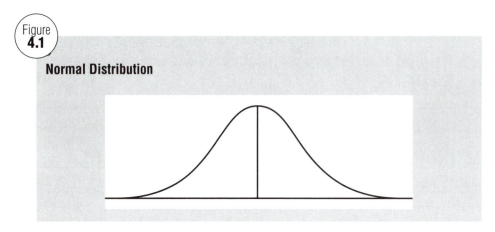

Figure 4.1

Normal Distribution

Check for Understanding

4.1 From the data presented in Table 4.1 (p. 51), calculate the percentile rank (PR) of the student with a 3.6023 GPA.

4.2 What is the percentile rank of the student in the 73rd rank position?

4.3 Define percentile rank.

Z-SCORES AND PERCENTILE RANKS

A *standard score* is a score that shows the relative standing, or the exact location, of a raw score in a distribution. For example, a student's SAT score is a standard score, because it shows how well the student scored in relation to other students who took the SAT. A standard score can be computed if the mean and the standard deviation of a population

distribution are known. If both are known and a very large population exists, the distribution will be very close to a perfectly symmetrical bell-shaped curve, which is also known as a normal curve or a normal distribution (see Chapter 2).

Although normal distributions are shaped alike and all are symmetrical, the scales for different tests are sometimes different. For example, the Wechsler IQ scale has a population mean (μ) = 100 and a population standard deviation (s) = 15, while the Stanford-Binet IQ scale has a population mean (μ) = 100 and a population standard deviation (s) = 16. Yet, even when the reported scales are slightly different, we can transform their data into standard deviation units by converting the scores into z-scores, which are standard scores representing the number of standard deviation units a raw score is above or below the mean. Equation 4.2 shows the formula for calculating a z-score.

Equation 4.2

$$z = \frac{x - \mu}{\sigma}$$

In equation 4.2, μ is the population mean, σ is the population standard deviation, and x is the raw score. The individual results of all norm-referenced standardized tests are based on a population mean and a population standard deviation; not the sample statistics that we discussed in previous chapters.

So what does a z-score tell us? The z-score tells us how many standard deviation units the raw score is above or below the mean (see Figure 4.3, p. 57). Notice that a z-score of 0 is directly in the center of the normal distribution. A positive z-score of 1 is one standard deviation above the mean; a z-score of 2 is two standard deviations above the mean; a z-score of 3 is three standard deviations above the mean. Similarly, on the opposite side of the central point, a z-score of –1 is one standard deviation unit below the mean; a z-score of –2 is two standard deviation units below the mean; a z-score of –3 is three standard deviation units below the mean.

As our first application of z-scores, let's examine some results from the SAT. The Educational Testing Service (ETS) provides information on the interpretation of test results on its website at *http://professionals.collegeboard.com/gateway* for test administrations between 2001 and 2002 before the new, three-part SAT Reasoning Test was introduced. For the 1,276,320 test takers during that time span the population mean (μ) is 1020 for the combined verbal and mathematics sections, with a population standard deviation (σ) of 208 on the combined verbal and mathematics sections.

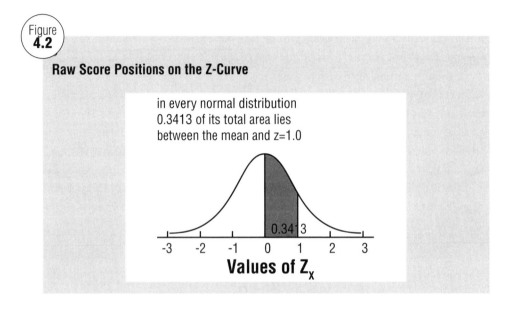

Figure 4.2

Raw Score Positions on the Z-Curve

in every normal distribution 0.3413 of its total area lies between the mean and z=1.0

Values of Z_x

Knowing the mean and standard deviation, we are now able to calculate a z-score. If a student received a combined raw score of $x = 1060$ on the SAT, what would his or her z-score be? Calculation 4.4 computes a z-score using Equation 4.2 (p. 55).

Calculation 4.4

$$z = \frac{1060 - 1020}{208} = 0.192$$

The calculated z-score of 0.192 gives us some very important information. First, because the score is positive, we know that the score is located somewhere above the mean (see Figure 4.3). This is, however, only an estimate of the exact location of the score with respect to the scores of the other test takers. To find the exact location, or percentile rank, relative to the position of the other test takers, we must go to the Appendix, Areas of the Standard Normal Distribution, which demonstrates how to calculate percentile rank.

To use this table, first go down the left column to match the first decimal place. Because we are matching a z-score of 0.192, you need to find 0.1, under "z" in the left column. Next, match the second decimal place with the corresponding numerical value, which is 0.09. Going down the 0.09 column to the row containing the first decimal places, the number that you should find is 0.0753. This is because we know that a normal curve is symmetrical, and the amount of area from the left tail to the middle point, or mean, equals 50% of the total area in the normal curve. The positive z-score that you just found is 0.0753, and this tells us the amount of area between the mean and our raw score. Cal-

CHAPTER 4

Z-Scores and the Normal Distribution

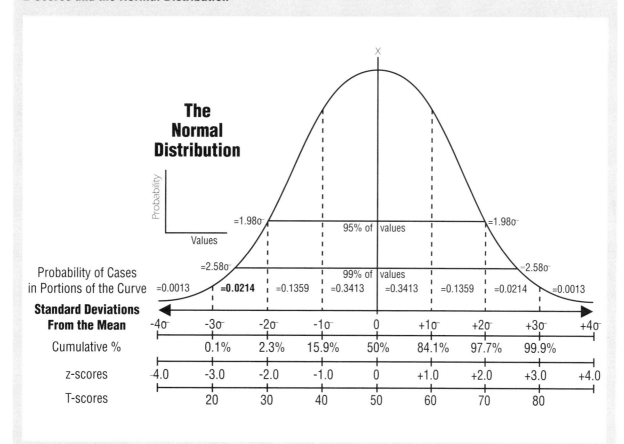

The Big Test

Nicholas Lemann's The Big Test (2000) gives a history of the SAT and how it affected socioeconomic status in America. The book explains the SAT's impact on the construction of an American meritocracy, the system of reward based on merit. According to Lemann, James Bryant Conant and Henry Chauncey designed the SAT so that the brightest members of society, especially members of the underprivileged classes, could gain access to higher education. As a result, a new class of well-educated people emerged, and formerly disadvantaged groups entered the middle and upper socioeconomic levels of society. Lemann contends that, because of the SAT, America has shifted from an aristocracy-based social system to a merit-based social system. Ironically, however, many opponents of standardized testing argue that the SAT has become the gatekeeper for entry into the most prestigious colleges and universities in America. As a result, critics suggest that a high percentage of working class and minority students, who have traditionally not fared well on the SAT, have been denied upward social mobility because of low SAT scores.

culation 4.5 shows how to calculate the percentile rank for a positive z-score by using the numerical value found in the table.

Calculation 4.5

0.5000 + 0.0753 = 0.5753
0.5753 × 100 = 57.53rd PR

Notice that in Calculation 4.5, 0.5000 represents the 50th percentile of area on the normal curve between the left tail and the position of the mean. The value of 0.0753 is the area between the mean and the exact location of the z-score. For example, when we add 0.5000 to 0.0753, we calculate a proportion of area that equals 0.5753 of the area within the normal curve. Moreover, to convert the proportion to a percentile rank, we must multiply the proportion by 100, which gives us a whole number. In this case, our z-score of 0.192 tells us that the SAT score of 1060 is at the 57.53rd percentile rank (PR). In other words, 57.53% of the test takers scored at or below the SAT raw score of 1060, which corresponds with the percentile rank data provided to test takers by ETS.

To determine the relative standing required for a student with an SAT raw score of 820, we can calculate a z-score and compute a percentile rank. The first step is to use Equation 4.2 (p. 55) to calculate the z-score. Calculation 4.6 shows the z-score calculation for an SAT score of 820.

Calculation 4.6

$$z = \frac{820 - 1020}{208} = -0.962$$

Remember, when calculating z-scores, you must know the population mean and standard deviation. Next, as in the previous z-score example in Calculation 4.5, we go to the Appendix, Areas of the Standard Normal Distribution, which will be used to compute a percentile rank. Again, take the value of the z-score, –0.96, and go down the left column to match the first decimal place, which is 0.09. By going across the row to 0.06, you should have found the numerical area proportion value of 0.3315 on the table. However, since the z-score is negative, you will need to subtract the proportion value from 0.500 rather than add. Calculation 4.7 shows how to calculate a percentile rank from a negative z-score by using the proportion value of 0.3315.

Calculation 4.7

0.5000 − 0.3315 = 0.1685
0.1685 × 100 = 16.85th PR

Since a negative z-score is below the mean, we subtract the area difference between the position of the raw score and the mean. On the basis of our calculation, the SAT raw score of 820 is equivalent to a z-score of −0.96 and a relative standing at the 16.85th PR. For those of you who may have taken the SAT before 1995, your scores have been recentered. Table 4.2 (p. 60) show how scores at the University of Virginia were recentered.

Check for Understanding

4.4. Using Table 4.2, what is the recentered SAT total for a student who had an SAT total of 1220 in 1984?
4.5. If the population mean (μ) is 100 for the Stanford-Binet Intelligence Test, with a population standard deviation (σ) of 16, what are the z-score and percentile rank for a person who scored 110?
4.6. Calculate a Stanford-Binet raw score (x) from a z-score of −1.25, knowing that the population mean (μ) is 100, with a population standard deviation (σ) of 16.

Z-SCORES, PERCENTILE RANK, AND IQ SCORES

Because public school teachers work with a variety of students who span a spectrum of cognitive abilities, it is important to discuss relative standing as it relates to intelligence quotient (IQ) scores. Table 4.3 (p. 61) shows the range of IQ classifications, displaying ranges from 65 and below to 128 and above. In general, teachers should understand that a typical class of students is composed of students who range in reported IQ from about 85 to 130, according to the Wechsler IQ Scale. In this case, what percentage of students from a normal population does this include? The first step to solve this problem is to calculate the z-score for a raw score of 85 (see Calculation 4.8). Next, calculate a z-score for a raw score of 130 (see Calculation 4.9).

Calculation 4.8

$$z = \frac{85 - 100}{15} = -1.00$$

Calculation 4.9

$$z = \frac{130 - 100}{15} = 2.00$$

CHAPTER 4

University of Virginia Recentered SAT Scores

Year	Non-recentered Scores			Recentered Scores[1]		
	SAT Verbal	SAT Math	SAT Total	SAT Verbal	SAT Math	SAT Total[2]
2007				645	662	1307
2006				654	671	1325
2005				653	667	1320
2004				659	671	1330
2003				654	670	1324
2002				647	668	1314
2001				648	665	1314
2000				643	661	1304
1999				648	659	1308

Year	Non-recentered Scores			Recentered Scores[1]		
	SAT Verbal	SAT Math	SAT Total	SAT Verbal	SAT Math	SAT Total
1989	577	641	1218	646	644	1290
1988	575	639	1214	645	642	1287
1987	585	645	1230	653	648	1301
1986	582	641	1223	650	644	1294
1985	586	635	1221	654	637	1291
1984	584	636	1220	652	638	1290
1983	579	623	1202	647	627	1274
1982	578	620	1198	647	625	1272
1981	567	607	1174	637	613	1250
1980	570	608	1178	639	613	1252
1979	584	615	1199	652	621	1273
1978	583	625	1208	652	629	1281
1977	588	620	1208	656	624	1280

[1] In April 1995, Educational Testing Service began recentering SAT scores so that the national mean scores for verbal and math would both be very close to 500. This caused mean verbal scores at UVa to increase by approximately 70 points, but had little impact on mean math scores at UVa. Although recentered scores did not exist for classes that entered before 1996, they have been calculated and reported above so scores for the years 1995 and earlier can be compared to the recentered scores.

[2] In some cases the mean recentered SAT total score does not equal the sum of the mean verbal plus the mean math score because the verbal and math means have been rounded to the nearest integer.

Note: Figures do not include transfer students.

Source: www.web.virginia.edu/iaas/data_catalog/institutional/data_digest/adm_total.htm

After calculating a z-score of −1.00 for a raw IQ score of 85, along with the z-score of 2.00 for an IQ score of 130, we must then compute an area interval to determine the percentage of students falling within this range (See Figure 4.3, p. 57). To do this, we must first find the area from the z-score of −1.00 to the center of the normal distribution, where the z-score is 0.00 (The z-score of 0 is in the same position as the mean). You can see the numerical area proportion associated with the interval value of 0.3413 in the Appendix, Areas of the Standard Normal Distribution.

Next, we need to find the area from the z-score of 2.00 to the center of the normal distribution, where the z-score is 0.00. (Again, the z-score of 0 is in the same position as the mean.) This is also seen as the numerical area proportion associated with the interval value of 0.4772 in the Appendix, Areas of the Standard Normal Distribution.

Now that we have calculated the area intervals, we need to add them together to compute the total area. Calculation 4.10 shows how to calculate an area interval proportion into a percentage.

Calculation 4.10

0.3413 + 0.4772 = 0.8185
0.8185 × 100 = 81.85%

We know that 81.85% of the population have IQ ranges between 85 and 130. Undoubtedly, this teaching dynamic creates a complicated state of affairs for the classroom teacher, given that the students in a typical classroom can range in abilities from "dull normal" to "very superior" (See Table 4.3). Moreover, this statistic also shows us that special educators, for the most part, work with about 18.15% of the student population, assuming that special education participation is based on cognitive abilities (some children have vision problems, hearing problems, or other problems, which often are unrelated to cognitive abilities).

Table 4.3

IQ Classifications and Percentile Rank

Classification	IQ Limits	Percent Included
Very Superior	128 and over	2.2
Superior	120–127	6.7
Bright Normal	111–119	16.1
Average	91–110	50
Dull Normal	80–90	16.1
Borderline	66–79	6.7
Defective	65 and below	2.2

Interpreting ITBS Scores

A raw score is the number of items that a student answers correctly on a test. For example, if Justin's raw score is 7 on the mathematics section of the Iowa Test of Basic Skills (ITBS) test and his raw score is 10 on the science section of the ITBS, we cannot conclude that his achievement is the same in mathematics and science. Therefore, raw scores are usually converted to standard scores (SS) or percentile ranks (PR).

A standard score (SS) is a numerical value that locates a student's academic achievement on a standard scale. With the ITBS, standard scores are based on the median performance of students during the spring of each academic year. Therefore, a score of 150 for a first-grade student indicates that this student is at the median level for all first graders who took this test. A standard score of 150 is the median performance of students in the spring of grade one. For eighth graders, the median performance in the spring is a standard score of 250.

Grade	1	2	3	4	5	6	7	8
SS	150	168	185	200	214	227	239	250

With the ITBS, a student's percentile rank can vary, depending on which group is used to determine the ranking. For example, a student is simultaneously a member of many different groups— among them being all students in his or her classroom, building, school district, state, and the nation. Different sets of percentile ranks permit schools to make the most relevant comparisons involving their students.

Check for Understanding

4.7. Based on the IQ classifications in Table 4.3 (p. 61), how would a child with an IQ of 110 be classified?

4.8. Compute a z-score and percentile rank for a student who has an IQ of 97. Assuming that we are using the Wechsler scale, what is this student's IQ classification according to Table 4.3?

4.9. What percentage of students have IQ scores over 130 on the Wechsler Scale? What would students with IQs exceeding 130 be classified as according to Table 4.3?

T-SCORES

A *T-score* is an alternative to the z-score that uses a mean of 50 as its central point, with a standard deviation of 10. All z-scores can be converted to T-scores. The formula for converting a z-score to a T-score is found in Equation 4.3. A T-score converts a z-score to a 100-point scale. This system is preferable because T-scores produce only positive integers, whereas z-scores can be reported as negative. For example, if a boy's standardized test score is reported as $z = -0.50$, the T-score equivalent is 45.

Equation 4.3

$$T = 50 + (10 \times \text{z-score})$$

As an example, the T-score that is equivalent to a z-score of 2.2 is computed in Calculation 4.11.

Calculation 4.11

$$\text{T-score} = 50 + (10 \times 2.20) = 72$$

Figure 4.3 (p. 57) shows the relationships among z-scores and T-scores. Notice that each scale has a common feature in that the original scores can be located in relation to the mean and standard deviation unit. One advantage of working with T-scores is that all scores can be put on a standard scale without the confusion of having to work with negative or obscure numbers.

Check for Understanding

4.10. What T-score is equivalent to a z-score of –1.50?
4.11. What T-score is equivalent to an SAT score of 1050?
4.12. Knowing that a T-score is 40, calculate the corresponding z-score.

Challenge Question

4.13. Given a normal distribution with a population mean of 100 and a standard deviation of 15, find the percentile ranks for the following raw scores:
   a. 145
   b. 130
   c. 115
   d. 100
   e. 85
   f. 70

SUMMARY

A *percentile rank* is the percentage of students whose scores fall at or below a particular score. Percentile ranks are computed by calculating a standard score called a z-score, which can be used to translate an individual's performance within a group. Percentile ranks are used by teachers to report a student's relative position among a group of students in terms of those students who are at or below that student's level of achievement or aptitude. Standardized tests, such as the SAT, Wechsler IQ Test, and the Iowa Tests of Basic Skills (ITBS),

are reported as norm-referenced population data. Therefore, scores from these tests can be converted into standard scores, such as z-scores and T-scores, and can be used to compute a student's relative standing as percentile rank. Teachers can use relative standing to report progress and gauge the effectiveness of their own teaching.

CALCULATOR EXPLORATION

Using the TI-83 or TI-84 graphing calculator, you can easily compute the percentage of students by area by following these key strokes.

Problem: Determine the percentage of students in a normal distribution who fall between an IQ range of 85 and 130. Since an 85 IQ is 1 standard deviation unit below the mean (equivalent to a z-score of –1.00) and an IQ of 130 is 2 standard deviation units above the mean (equivalent to a z-score of 2.00), we will use the TI-83 graphing calculator to determine the percentage of students in a normal distribution who fall within this range of scores.

TI-83/TI-84 Calculator

Step 1. First press [2nd VARS] to activate the distribution function.
Next, select [2: normalcdf(] by pressing 2.

Step 2. Next, enter the corresponding lower z-score of –1, insert a comma, and enter the higher z-score of +2 and close the parenthesis. Next press [ENTER].

Note: The calculator has computed the proportion of area in a normal curve that corresponds with this range of scores. To get the percentage, you should multiply the proportion by 100. Thus, going out four decimal places (0.8185 × 100 = 81.85%) is a much easier way to determine the area than the table that we used earlier in the chapter.

CHAPTER REVIEW QUESTIONS

4.14. Of the following z-scores, which value indicates the greatest numerical distance from the mean?
a. z = −1.00
b. z = +1.75
c. z = −2.75
d. z = +2.50

4.15. If the Wechsler IQ scale has a population mean (μ) = 100 and a population standard deviation (s) = 15, what is the z-score for a student scoring 80 on the Wechsler IQ test?

4.16. Based on the z-score calculated in question 4.15, what is the percentile rank of a student who scores 80 on the Wechsler IQ test?

4.17. If the Wechsler IQ scale has a population mean (μ) = 100 and a population standard deviation (s) = 15, what is the z-score for a student scoring 120 on the Wechsler IQ test?

4.18. Based on the z-score calculated in question 4.17, what is the percentile rank of a student who scores 120 on the Wechsler IQ test?

4.19. If the Stanford-Binet IQ scale has a population mean (μ) =100 and a population standard deviation (s) = 16, what is the z-score for a student scoring 80 on the Stanford-Binet IQ test?

4.20. Based on the z-score calculated in question 4.19, what is the percentile rank of a student who scores 80 on the Stanford-Binet IQ test?

4.21. If the Stanford-Binet IQ scale has a population mean (μ) =100 and a population standard deviation (s) = 16, what is the z-score for a student scoring 120 on the Stanford-BinetIQ test?

4.22. Based on the z-score calculated in question 4.21, what is the percentile rank of a student who scores 120 on the Stanford-Binet IQ test?

4.23. The population mean (μ) is 1020 for the combined verbal and mathematics sections of the SAT, with a population standard deviation (s) of 208 on the combined verbal and mathematics sections. Calculate the following:
a. A z-score for a combined SAT score of 1270
b. A percentile rank for a combined SAT score of 1270
c. A T-score for for a combined SAT score of 1270
d. A z-score for a combined SAT score of 770
e. A percentile rank for a combined SAT score of 770
f. A T-score for for a combined SAT score of 770

4.24. Explain the meaning of percentile ranks and recentered scores.

ANSWERS: CHECK FOR UNDERSTANDING

4.1. 75.5th percentile rank
4.2. 27.5th percentile rank
4.3. A percentile rank is the percentage of scores that falls below a given score. Sometimes the percentage is defined to include all scores that fall at the point; sometimes the percentage is defined to include half of the scores at the point.
4.4. 1290 4.5. z-score = 0.625, percentile rank = 73.57th PR
4.6. The subject would have an IQ raw score of 80.
4.7. This student is classified as average.
4.8 . The z-score is –0.20, with a relative standing at the 42.07th PR. This student is classified as average, according to Table 4.3.
4.9. 2.28% of all students. All students with IQs above 130 fall into Table 4.3's very superior classification.
4.10. T-score is 35.
4.11. T-score is 51.44.
4.12. z-score equals –1.00.

ANSWER: CHALLENGE QUESTION

4.13. a. 145 = 99.87th PR
b. 130 = 97.72nd PR
c. 115 = 84.13rd PR
d. 100 = 50th PR
e. 85 = 15.87th PR
f. 70 = 2.28th PR

ANSWERS: CHAPTER REVIEW QUESTIONS

4.14. c. z = –2.75
4.15. z = –1.33
4.16. PR = 9.18th
4.17. z = +1.33
4.18. PR = 90.82nd
4.19. z = –1.25
4.20. PR = 10.56th
4.21. z = +1.25
4.22. PR = 89.44th
4.23. a. z = 1.20
b. 88.49th PR

c. T = 38
d. z = –1.20
e. 11.51st PR
f. T = 36.80

4.24. Percentile ranks are the relative standing of a score in a distribution of scores. The percentile rank is the relative standing at or below the raw score. Recentering of scores occurs when variables such as the mean and standard deviation are changed. For example, in April 1995, Educational Testing Service began recentering SAT scores so that the national mean scores for verbal and math would both be very close to 500.

INTERNET RESOURCES

This part of the Teachers and Families website gives a general overview of percentiles and standard scores. The site is designed for someone who has a minimal understanding of the interpretation of test scores. It is a useful resource for practicing teachers to use as a reference, as well as a good site to suggest to parents who have test interpretation questions. *www.teachersandfamilies.com/open/parent/scores2.cfm*

Because the Iowa Tests of Basic Skills (ITBS) is used throughout the nation from grades K through 8, this site can be used by elementary and middle school teachers as a model depicting an essential standardized achievement test. *www.education.uiowa.edu/itp/itbs/itbs_interp_score.htm*

This site, by Professor Gary McClelland of the University of Colorado, offers a z-score calculator and the corresponding probabilities, which are equivalent to area calculations. This is a handy resource for students looking for additional information about z-scores. *http://psych.colorado.edu/~mcclella/java/normal/normz.html*

REFERENCE

Lemann, N. 2000. *The big test: The secret history of the American meritocracy*. New York: Farrar, Straus, and Giroux.

FURTHER READING

American Educational Research Association (AERA). 1999. *Standards for educational and psychological testing*. Washington, DC: American Educational Research Association.

Brown, J. R. 1991. The retrograde motion of planets and children: Interpreting percentile rank. *Psychology in the Schools* 28 (4): 345–353.

Journal of School Improvement. 2000. What is this standard score stuff, anyway? *Journal of School Improvement* 1 (2): 44–45.

Pearson, K. 1930. *Life, letters, and labours of Francis Galton*. Vol. IIIa, *Correlation, personal identification, and eugenics*. Cambridge, England: Cambridge University Press.

Raymondo, J. C. 1999. *Statistical analysis in the behavioral sciences*. New York: McGraw Hill.

Thorne, B. M., and J. M. Giesen, 2000. *Statistics for the behavioral sciences*. Mountain View, CA: Mayfield.

Zawojewski, J. S., and J. M. Shaughnessy. 2000. Mean and median: Are they really so easy? *Mathematics Teaching in the Middle School* 5 (7): 436–440.

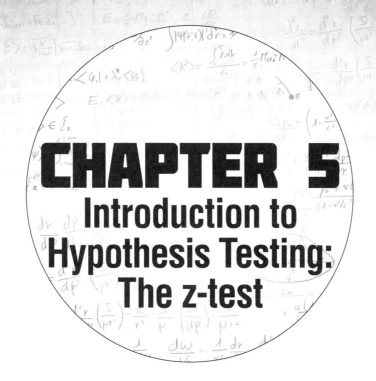

CHAPTER 5
Introduction to Hypothesis Testing: The z-test

OBJECTIVES

When you complete this chapter, you should be able to
1. demonstrate an understanding of the z-test and its relevance to classroom settings;
2. compare the relationship between descriptive statistics and parametric procedures;
3. calculate z-tests from classroom and standardized test results; and
4. find, use, and interpret z-test results.

Key Terms

When you complete this chapter, you should understand the following terms:
    parametric
    inferential statistics
    significant
    null hypothesis
    alternative hypothesis
    z-test
    standard error of the mean
    one-tailed test
    two-tailed test
    type-I error

INTRODUCTION

Statistical hypothesis testing is a methodology that uses experimental data for decision making. In statistical analysis, the end result is either a *significant* or *not significant* finding. When a finding is significant, the researcher is confident that the significant finding most likely did not occur by chance. Hence, when Sir Ronald Fisher discussed the difference between means, the term *significance test* was coined (Fisher et. al. 1990). Hypothesis testing confirms data analysis findings through a probability figure linked to a null-hypothesis rejection. This probability is the possibility of the researcher committing a type-I error (a false positive), meaning a null hypothesis that is rejected is true.

With the hypothesis test, the critical region is the range of values that are located at the tail or tails of the normal curve distribution. Therefore, the calculated value of the test is compared to the critical value to determine whether to reject or to not reject the null-hypothesis. Based on the probability value selected by the researcher (e.g., 0.05 or 0.01), a statistical table is used to determine the critical value. For example, if a researcher sets the alpha level at 0.05 for a two-tailed test, the critical value (C.V.) is +/– 1.96. However, if the alpha level is set at 0.05 for a one-tailed test, the critical value (C.V.) is 1.65 and can be positive or negative depending on the direction of the hypothesis.

THE COURT ROOM TRIAL EXAMPLE

Similar to a court room trial, the statistical test procedure helps make decisions where the burden of proof is required. For example, you may have heard "innocent until proven guilty in the court of law." Like the prosecuting attorney attempting to prove the guilt of the defendant, a researcher uses a statistical test to analyze the data and computes a statistical test. The statistical test result is compared to a critical value (C.V.), which is based on probability value (p).

When comparing the statistic to the C.V., if the statistic exceeds the C.V., the researcher rejects the null hypothesis. However, there is a probability of a type-I error (type-I error, meaning the null hypothesis was rejected and should not have been rejected.) and the level of probability is equivalent to the p-value, which is set by the researcher prior to the statistical test (priori). For example, if the researcher sets the p-value at 0.05, the probability of committing a type-I error is 5% or five times out of 100. Using the legal case, Table 5.1 shows how a defendant can be found innocent or guilty based on this logic.

The z-test (z) is a test statistic for which the normal distribution is used for the statistic to test the null hypothesis. Typically, this statistical test is used when small sample sizes are available to generalize findings to a population, and is sometimes known as an approximate test.

Historically, it seems that John Arbuthnot (1667–1735), submitted the earliest example of a statistical test when he published *An Argument for Divine Providence, Taken From the Constant Regularity Observed in the Births of Both Sexes* in 1710, while physician to Queen Anne. This paper discussed the difference between male births and female births. Since no statistical significance was found, Arbuthnot argued that divine providence, not probability, determines the gender ratio at birth (Glass et al. 2008).

Table 5.1

Type-I Error and the Null Hypothesis

	Null Hypothesis (H_0) is true. He truly is innocent.	Alternative Hypothesis (H_1) is true. He truly is guilty.
Accept Null Hypothesis	Right decision	Wrong decision Type-II Error
Reject Null Hypothesis	Wrong decision Type-I Error	Right decision

THE ONE-SAMPLE Z-TEST

Equations:

$$SE = \frac{\sigma}{\sqrt{n}}$$

$$Z = \frac{\bar{x} - \mu}{SE}$$

The one-sample z-test statistic is a ***parametric*** statistic, which is a branch of statistics that assumes that data have come from a type of normal distribution and makes inferences about the parameters of the distribution. Therefore, parametric procedures will evaluate type-I error by using an alpha level to determine a critical value from a z table found from a normal distribution and compare it to the statistical computation.

It is important to know how to apply the z-test to research because often data are from interval and ratio scales of measurement. As discussed earlier, interval and ratio scales are classifications of numbers in equal units. In education, school districts categorize students based on ability levels (i.e., general education and gifted education). Hence, a teacher might find that students are better at mathematics computations when comparing the ability levels of students in a gifted and talented physics course to those in the general population.

THE ONE-SAMPLE Z-TEST (TWO-TAILED TEST)

To illustrate an application of the one-sample z-test, we will compare the IQ scores of students who are enrolled in a gifted and talented program to those in the general population.

Example 1

Table 1. Comparison of IQ Scores Between a Local School's Gifted and Talented Program to Those in the General Population

General Population IQ Scores	Gifted and Talented IQ Scores
Population Mean (μ) = 100	Sample Mean ($\bar{x}$) = 130
Population Standard Deviation (σ) = 15	Sample size (n) = 20

To begin the calculation, we must follow a sequence of steps that will be used throughout hypothesis testing. As you will see, the most general way to compute a z-test is to define the statistical test statistic that can be calculated from scale of measurement and the collection of data. Below are the Seven Steps for Hypothesis Testing, which will be used in all of the statistical tests used throughout this book.

Seven Steps for Hypothesis Testing

STEP 1: Identify the appropriate statistical test to be used with the given data for analysis.
STEP 2: Hypotheses • H_o (Null Hypothesis): This hypothesis proposes that no statistical significance exists. The null hypothesis attempts to show that no variation exists between the means, or that a single variable is not different. When the null hypothesis is not rejected, it is presumed to be true until statistical evidence nullifies it for an alternative hypothesis. • H_1 (Alternative Hypothesis): The alternative hypothesis is the hypothesis that is contrary to the null hypothesis. This hypothesis should correspond with the researcher's assumption and is used to discern the conclusion.
STEP 3: Alpha Level (α) Alpha level is the probability of making a type-I error (rejecting the null hypothesis when the null hypothesis is true). The lower this value is, the less chance that the researcher has of committing a type-I error. Also, an experiment should use a sample size large enough (i.e., n = 20) so that there is adequate statistical power. As a general rule, statistical analysis in education uses alpha levels of 0.05 or sometimes 0.01.
STEP 4: Critical Value (C.V.) A number that causes rejection of the null hypothesis if a given test statistic is this number or more, and nonrejection of the null hypothesis if the test statistic is smaller than this number. The critical value can be found by analyzing a statistical table related to the appropriate statistical test.

Continued on the next page.

Continued from previous page.

> **STEP 5: Statistical Test**
> A statistical test provides a mechanism for making quantitative decisions. The intent is to determine whether there is enough evidence to "reject" the null hypothesis. Not rejecting may be a good result if we want to continue to act as if we "believe" the null hypothesis is true. Or it may be a disappointing result, possibly indicating we may not yet have enough data to "prove" something by rejecting the null hypothesis.
>
> **STEP 6: Reject or Do Not Reject the Null Hypothesis**
> By comparing the critical value to the calculated statistics, if the statistic exceeds the critical value, the null hypothesis is rejected.
>
> **STEP 7: Interpretation**
> Rejecting the null hypothesis may disclose that an independent variable/treatment did affect the dependent variable. However, not rejecting the Null Hypothesis may be a good result if we are looking for a result where no difference is desirable. Or it may be a disappointing result, possibly indicating we may not yet have enough data to "prove" something by rejecting the null hypothesis.

After understanding that the data must be analyzed by using a z-test, we can proceed with the hypothesis.

> **STEP 1: Identify the Statistical Test**
> z-test

Most researchers test the null hypothesis (H_0), which is a statement that there is "no difference" between the sample mean and the population mean. Conversely, the alternative hypothesis (H_1) indicates that the sample mean and the population mean are not equal. In this case the hypotheses are written.

> **STEP 2: Hypotheses**
> H_0: There is no mean difference between the IQs of students at a local school and the IQs of students within the population.
> or
> H_0: $(\mu) = 100$
> H_1: There is a mean difference between the IQs of students at a local school and the IQs of students within the population.
> or
> H_1: $(\mu) \neq 100$

The third step is to set the alpha level (α). As explained earlier, the alpha level is used to determine the probability of committing a type 1 error. Therefore, different studies have advantages or disadvantages when using higher or lower α-levels. Hence, you have to be careful about interpreting the meaning of these terms. For example, *higher* α-levels can *increase* the chance of a type I error. Therefore, a lower α-level indicates that you are conducting a test that is more rigorous. Keep in mind that an α of 0.05 means that the researcher is taking a risk of being wrong 5 in 100 times when rejecting the null hypothesis (e.g., Committing a Type I error is saying there's an effect when there really is not.). However, an alpha level (α) of 0.01 (compared with 0.05) means the researcher is being relatively cautious by minimizing the risk of being wrong only 1 in 100 times when rejecting the null hypothesis. However, the researcher decides the alpha level α based on the risk. For example, the safety of medicine use requires more scrutiny than substituting a calculator for microcomputer-based software for calculations. Therefore, in this case, the alpha level α is set at 0.05.

> **STEP 3: Alpha Level**
> α = 0.05

The fourth step is to determine the critical value. The Critical Values table for the z-score is used for the z-test and is used to establish the critical value. First, you need to determine the direction of the statistical test or whether to use a one-tail or a two-tailed test. The two-tailed test is used in inferential statistics when H_0 (the null hypothesis) could be rejected when the value of the test statistic is either sufficiently small or sufficiently large. This is in contrast to the one-tailed test because with the one-tailed test, only one rejection region is sufficiently small or sufficiently large and is preselected according to the direction of the alternative hypothesis and is rejected only if the test statistic satisfies this criteria.

In this case, since we have not specified any direction, we will look for a critical value for a two-tailed test. For the purpose here, we will only use critical values for alpha levels of 0.05, 0.01, or 0.001. Therefore, the critical values from the statistical table are:

Critical Values for z-test*

Alpha Level	One-Tail	Two-Tail
α = 0.05	C.V. = +/– 1.64	C.V. = 1.96/–1.96
α = 0.01	C.V. = +/– 2.33	C.V. = 2.58/–2.58
α = 0.001	C.V. = +/– 3.10	C.V. = 3.30/–3.30

*You will have only one critical value for a one-tail test and the value can be positive or negative. However, for the two-tailed test, there are two critical values listed, both positive and negative.

If other alpha levels are required, the critical value can be determined by examining the boundaries for the critical region based on the Areas of the Standard Normal Distribution in the Appendix. For an alpha level of 0.05, the critical value = 1.96/–1.96.

STEP 4: Critical Value
C.V. = 1.96/–1.96

The fifth step is to compute the z-test statistic. The formula for the z-test is:

$$SE = \frac{\sigma}{\sqrt{n}}$$

$$Z = \frac{\bar{x} - \mu}{SE}$$

In this equation, SE represents the Standard Error and estimates the amount of sampling error between the sample and the population. Since the population standard deviation (σ) = 15 and the sample size (n) = 20, the calculated SE = 3.355.

STEP 5: Statistical Test (z-test)
Step #1 =

$$SE = \frac{15}{\sqrt{20}} = 3.35$$

Step #2 =
Next we calculate the z-test, which is the distance from the sample mean to the population mean in units of the standard error:

$$Z = \frac{130 - 100}{3.355} = 8.94$$

The calculation for the z-test = 8.94, which measures the difference between the IQs of students at a local school and the IQs of students within the population. The sixth step is to reject or to not reject the null hypothesis. This step takes into consideration a comparison of the critical value to the calculated z-test statistic, which compares the z-test of 8.94 to the C.V. of 1.96/–1.96. Subsequently, since the z-test statistic exceeds the critical value of 1.96/–1.96, the null hypothesis is rejected.

STEP 6: Reject or Do Not Reject the Null Hypothesis
Since the z-test of 8.94 is larger than C.V. of 1.96/–1.96, we reject the null hypothesis.

The seventh step is to interpret our findings. Since we rejected our null hypothesis,

we conclude that the observed sample mean is significantly different from the population. Therefore, we conclude that the IQs of students enrolled in a gifted and talented program at a local school are significantly higher than the IQs of students within the population.

> **STEP 7: Interpretation**
> Rejecting the null hypothesis may disclose that an independent variable/treatment did affect the dependent variable. There is a significant difference between the IQs of students enrolled in a gifted and talented program at a local school and the IQs of students within the population.

Example 2

The One-Sample z-test (One-Tailed Test)

To illustrate an application of the one-sample z-test, we will compare the IQ scores of students who are enrolled in a gifted and talented program to those in the general population. In this case, however, the researcher specifies direction and states that students in gifted and talented programs "have higher" IQs than the general population.

Table 1. Comparison of IQ Scores Between a Local School's Gifted and Talented Program to Those in the General Population

General Population IQ Scores	Gifted and Talented IQ Scores
Population Mean (μ) = 100	Sample Mean ($\bar{x}$) = 130
Population Standard Deviation (σ) = 15	Sample size (n) = 20

> **STEP 1: Identify the Statistical Test**
>
> z-test

Most researchers test the null hypothesis (H_0), which is a statement that there is "no difference" between the sample mean and the population mean. Conversely, the alternative hypothesis (H_1) indicates that the sample mean and the population mean are not equal and that the sample is greater than the population. In this case the hypotheses are written.

> **STEP 2: Hypotheses**
> H_o: There is no mean difference between the IQs of gifted and talented students at a local school and the IQs of students within the population.
> or
> H_o: $(\mu) \leq 100$
> H_1: The IQs of gifted and talented students at a local school and the IQs of students within the population.
> or
> H_1: $(\mu) > 100$

The third step is to set the alpha level (α). As explained earlier, the alpha level is used to determine the probability of committing a type-I error. Therefore, different studies have advantages or disadvantages when using higher or lower α-levels. Based on the risk, the researcher decides the alpha level α. In this case, the alpha level α be set at 0.01.

> **STEP 3: Alpha Level**
> $\alpha = 0.01$

The fourth step is to determine the critical value. Since this is a one-tailed test (i.e., the research specified that we will look for a critical value for a z-test where students "have higher" IQs), we will use a one-tail test with an alpha level of 0.01. Again, for the purpose here, we will only use critical values for alpha levels of 0.05 or 0.01. Therefore, the critical values from the statistical table are in the table below:

Critical Values for z-test*

Alpha Level	One-Tail	Two-Tail
$\alpha = .05$	C.V. = +/−1.64	C.V. = 1.96/−1.96
$\alpha = .01$	C.V. = +/−2.33	C.V. = 2.58/−2.58
$\alpha = .001$	C.V. = +/−3.10	C.V. = 3.30/−3.30

*You will have only one critical value for a one-tail test and the value can be positive or negative. However, for the two-tailed test, there are two critical values listed, both positive and negative.

For an alpha level of .01, the Critical Value = +2.33.

> **STEP 4: Critical Value**
> C.V. = +2.33

The fifth step is to compute the z-test statistic. The formula for the z-test is:

$$SE = \frac{\sigma}{\sqrt{n}}$$

$$Z = \frac{\bar{x} - \mu}{SE}$$

In this equation, SE represents the Standard Error and estimates the amount of sampling error between the sample and the population. Since the population standard deviation (σ) = 15 and the sample size (n) = 20, the calculated SE = 3.35.

> **STEP 5: Statistical Test (z-test)**
> Step #1 =
>
> $$SE = \frac{15}{\sqrt{20}} = 3.35$$
>
> Step #2 =
> Next we calculate the z-test, which is the distance from the sample mean to the population mean in units of the standard error:
>
> $$Z = \frac{130 - 100}{3.355} = 8.94$$

Again, since we used the same data, the calculation for the z-test = 8.94, which measures the difference between the IQs of students at a local school and the IQs of students within the population. The sixth step is to reject or to not reject the null hypothesis. This step takes into consideration a comparison of the critical value to the calculated z-test statistic, which compares the z-test of +8.94 to the C.V. of +2.33. Subsequently, since the z-test statistic exceeds the critical value of +2.33, the null hypothesis is rejected.

> **STEP 6: Reject or Do Not Reject the Null Hypothesis**
> Since the z-test of +8.94 is larger than C.V. of +2.33, we reject the null hypothesis.

The seventh step is to interpret our findings. Since we rejected our null hypothesis, we conclude that the observed sample mean is significantly different from the population. Therefore, we conclude that the IQs of students enrolled in a gifted and talented program at a local school are significantly higher than the IQs of students within the population.

> **STEP 7: Interpretation**
> Rejecting the null hypothesis may disclose that an independent variable/treatment did affect the dependent variable.
> There is a significant difference between the IQs of students enrolled in a gifted and talented program at a local school and the IQs of students within the population and the students in the gifted and talented program are significantly higher than the general population.

CALCULATOR EXPLORATION

Using the TI-83 or TI-84, you can easily compute a z-test. Using the two-tailed test from Example 1, we will demonstrate how to calculate a z-test with a graphing calculator.

Press [**STAT**].
Select [**TESTS**] [**1:Z-Test…**]
Press [**ENTER**].

Highlight [**Stats**] and enter the population mean, population standard deviation, sample mean, and sample size. Also, make sure that the not equal symbol is highlighted and Press [**Calculate**].

Results

Graph with normal curve area is available if you select [**Draw**].

The z-test test statistic is +8.94 since +8.94 > +/− 1.96… and p-value is less than 0.05. Subsequently, we reject the null hypothesis.

SUMMARY

A statistical hypothesis test is an inferential procedure using data from samples to draw a general conclusion about a population. Once the data has been scrutinized, the appropriate statistical test can be identified and a hypothesis is rendered about the population. Subsequently, the sample provides data to make a decision to reject or to not reject the hypothesis.

Hypothesis testing was introduced by using the z-test, where a sample mean is used to test the hypothesis related to the population mean. Typically, the sample group receives a treatment (e.g., computer-assisted instruction versus traditional instruction) and the question is to determine whether or not the treatment has had an effect on the group of interest.

In this chapter, we structured hypothesis testing as a seven-step process that will be used throughout the remainder the inferential statistics used in this book.

Seven Steps for Hypothesis Testing

STEP 1: Identify the appropriate statistical test to be used with the given data for analysis.
STEP 2: Hypotheses • H_o (Null Hypothesis): This hypothesis proposes that no statistical significance exists. The null hypothesis attempts to show that no variation exists between the means, or that a single variable is not different. When the null hypothesis is not rejected, it is presumed to be true until statistical evidence nullifies it for an alternative hypothesis. • H_1 (Alternative Hypothesis): The alternative hypothesis is the hypothesis that is contrary to the null hypothesis. This hypothesis should correspond with the researcher's assumption and is used to discern the conclusion.
STEP 3: Alpha Level (α) Alpha level is the probability of making a type-I error (rejecting the null hypothesis when the null hypothesis is true). The lower that this value is, the less chance that the researcher has of committing a type-I error. Also, an experiment should use a sample size large enough (i.e., $n = 20$) so that there is adequate statistical power. As a general rule, statistical analysis in education uses alpha levels of 0.05 or sometimes 0.01.
STEP 4: Critical Value (C.V.) A number that causes rejection of the null hypothesis if a given test statistic is this number or more, and nonrejection of the null hypothesis if the test statistic is smaller than this number. The critical value can be found by analyzing a statistical table related to the appropriate statistical test.

Continued on next page.

Continued from previous page.

> **STEP 5: Statistical Test**
> A statistical test provides a mechanism for making quantitative decisions. The intent is to determine whether there is enough evidence to reject the null hypothesis. Not rejecting may be a good result if we want to continue to act as if we believe the null hypothesis is true. Or it may be a disappointing result, possibly indicating we may not yet have enough data to prove something by rejecting the null hypothesis.
>
> **STEP 6: Reject or Do Not Reject the Null Hypothesis**
> By comparing the critical value to the calculated statistics, if the statistic exceeds the critical value, the null hypothesis is rejected.
>
> **STEP 7: Interpretation**
> Rejecting the null hypothesis may disclose that an independent variable/treatment did affect the dependent variable. However, not rejecting the null hypothesis may be a good result if we are looking for a result where no difference is desirable. Or it may be a disappointing result, possibly indicating we may not yet have enough data to prove something by rejecting the null hypothesis.

The term *significance test* was first mentioned by Ronald Fisher, when he explored whether there is difference between two means. In hypothesis testing, according to a predetermined upper limit probability, a result is called statistically significant if it is unlikely to have occurred by chance alone. Hence, hypothesis testing is a fundamental technique for statistical inference and helps researchers use data to optimize decisions by minimizing the risk of type-I error (rejecting the null hypothesis when the null hypothesis is true).

Keep in mind that when a researcher thinks that a treatment will affect scores in a certain direction (an increase or a decrease), the research may choose a directional hypothesis (one-tail rather than two-tail). Next, the research will determine the power of the hypothesis test (alpha level), which is the probability of committing a type-I error. After the alpha level is selected, the critical value is selected from the z-distribution table (e.g., a one-tailed test with an alpha level of 0.001 gives a critical value of +/– 3.30). The next step is to calculate a z-test and compare the z-test to the critical value (C.V.). If the z-test exceeds the C.V., the null hypothesis is rejected and an interpretation follows.

CHAPTER REVIEW QUESTIONS: (SEVEN STEPS FOR HYPOTHESIS TESTING)

1. Determine whether or not the sample is statistically different from the population at a significance level of 0.05.
   $\bar{x} = 566$ $\sigma = 30$ $n = 36$ $\mu = 560$

ANSWER FOR QUESTION 1

Seven Steps for Hypothesis Testing

STEP 1: z-test
STEP 2: Hypotheses H_o: $(\mu) = 560$ H_1: $(\mu) \neq 560$
STEP 3: Alpha Level (α) Alpha level is 0.05
STEP 4: Critical Value (C.V.) C.V. = +/− 1.96
STEP 5: Statistical Test z = 1.20
STEP 6: Do Not Reject Null Hypothesis
STEP 7: Interpretation *Summary* There is no statistically significant mean difference between the sample and the population (z, (n = 36) = 1.20, p > 0.05).

2. Determine whether or not the sample is statistically different from the population at a significance level of 0.01.

   $\bar{x} = 24$ $\sigma = 4$ $n = 64$ $\mu = 25$

ANSWER FOR QUESTION 2

Seven Steps for Hypothesis Testing

STEP 1: z-test
STEP 2: Hypotheses $H_0: (\mu) = 25$ $H_1: (\mu) \neq 25$
STEP 3: Alpha Level (α)   Alpha level is 0.01
STEP 4: Critical Value (C.V.)   C.V. = +/− 2.58
STEP 5: Statistical Test $z = -2.00$
STEP 6: Do Not Reject Null Hypothesis
STEP 7: Interpretation   *Summary*   There is no statistically significant mean difference between the sample and the population (z, (n = 64) = −2.00, p > 0.01).

3. Determine whether or not the sample is statistically different from the population at a significance level of 0.05.

   $\bar{x} = 82 \quad \sigma = 14 \quad n = 49 \quad \mu = 75$

ANSWER FOR QUESTION 3

Seven Steps for Hypothesis Testing

STEP 1: z-test
STEP 2: Hypotheses H_0: $(\mu) = 75$ H_1: $(\mu) \neq 75$
STEP 3: Alpha Level (α) Alpha level is 0.05
STEP 4: Critical Value (C.V.) C.V. = +/− 1.96
STEP 5: Statistical Test z = 3.50
STEP 6: Reject Null Hypothesis
STEP 7: Interpretation
Summary There is a statistically significant mean difference between the sample and the population, where the sample mean is significantly high than the population mean (z, (n = 49) = 3.50, p < 0.05).

4. Determine whether or not the sample is statistically different from the population at a significance level of 0.05.

   $\bar{x} = 109$ $\sigma = 15$ $n = 106$ $\mu = 100$

CHAPTER 5

ANSWER FOR QUESTION 4

Seven Steps for Hypothesis Testing

STEP 1: z-test
STEP 2: Hypotheses H_o: $(\mu) = 100$ H_1: $(\mu) \neq 100$
STEP 3: Alpha Level (α)   Alpha level is 0.05
STEP 4: Critical Value (C.V.)   C.V. = +/− 1.96
STEP 5: Statistical Test   z = 6.18
STEP 6: Reject Null Hypothesis
STEP 7: Interpretation   *Summary*   There is a statistically significant mean difference between the sample and the population, where the sample mean is significantly high than the population mean (z, (n = 106) = 6.18, $p < 0.05$).

5. Determine whether or not the sample is statistically different from the population at a significance level of 0.001.

   $\bar{x} = 940$ $\sigma = 200$ n = 203 $\mu = 1000$

ANSWER FOR QUESTION 5
Seven Steps for Hypothesis Testing

STEP 1: z-test
STEP 2: Hypotheses H_o: $(\mu) = 1000$ H_1: $(\mu) \neq 1000$
STEP 3: Alpha Level (σ Alpha levels is 0.001.
STEP 4: Critical Value (C.V.) C.V. = +/− 3.10
STEP 5: Summary Statistical Test z = −4.27
STEP 6: Reject Null Hypothesis
STEP 7: Interpretation
Summary There is a statistically significant mean difference between the sample and the population, where the sample mean is significantly lower than the population mean (z, (n = 203) = −4.27, p < 0.001)

6. Determine whether or not the sample is statistically higher than the population at a significance level of 0.05.
   $\bar{x} = 1100$ $\sigma = 200$ n = 10 $\mu = 1000$

ANSWER FOR QUESTION 6

Seven Steps for Hypothesis Testing

STEP 1: z-test
STEP 2: Hypotheses $H_o: (\mu) \leq 1000$ $H_1: (\mu) > 1000$
STEP 3: Alpha Level (α)   Alpha level is 0.05.
STEP 4: Critical Value (C.V.)   C.V. = +1.64
STEP 5: Statistical Test   z = 5.02
STEP 6: Reject Null Hypothesis
STEP 7: Interpretation   *Summary*   There is a statistically significant mean difference between the sample and the population, where the sample mean is significantly higher than the population mean (z, (n = 101) = 5.02, p < 0.05).

7. Determine whether or not the sample is statistically higher than the population at a significance level of 0.01.
    $\bar{x} = 75 \quad \sigma = 2.58 \quad n = 20 \quad \mu = 68$

ANSWER FOR QUESTION 7

Seven Steps for Hypothesis Testing

STEP 1: z-test
STEP 2: Hypotheses H_o: $(\mu) \leq 68$ H_1: $(\mu) > 68$
STEP 3: Alpha Level (α)   Alpha level is 0.01.
STEP 4: Critical Value (C.V.)   C.V. = +2.33
STEP 5: Statistical Test   z = 12.13
STEP 6: Reject Null Hypothesis
STEP 7: Interpretation   *Summary*   There is a statistically significant mean difference between the sample and the population, where the sample mean is significantly higher than the population mean (z, (n = 20) = 12.13, p < 0.01).

8. Determine whether or not the sample is statistically lower than the population at a significance level of 0.001.

   $\bar{x} = 77$ $\sigma = 10$ $n = 25$ $\mu = 100$

ANSWER FOR QUESTION 8

Seven Steps for Hypothesis Testing

STEP 1: z-test
STEP 2: Hypotheses H_o: $(\mu) \leq 100$ H_1: $(\mu) < 100$
STEP 3: Alpha Level (α) Alpha level is 0.001.
STEP 4: Critical Value (C.V.) – C.V. = –3.10
STEP 5: Statistical Test z = –11.50
STEP 6: Reject Null Hypothesis
STEP 7: Interpretation
Summary There is a statistically significant mean difference between the sample and the population, where the sample mean is significantly lower than the population mean (z, (n = 25) = –11.50, p < 0.001).

9. For the population at large, the Wechsler Intelligence Scale has a mean of 100 and a standard deviation of 15. School district officials wonder whether, on average, their students correspond to the national average. A random sample of 25 students produces a mean of 105. Test the null hypothesis at a significance level of 0.01.

ANSWER FOR QUESTION 9

Seven Steps for Hypothesis Testing

STEP 1: z-test
STEP 2: Hypotheses H_o: $(\mu) = 100$ H_1: $(\mu) \neq 100$
STEP 3: Alpha Level (α)   Alpha level is 0.01.
STEP 4: Critical Value (C.V.)   C.V. = +/− 2.58
STEP 5: Statistical Test   z = 1.66
STEP 6: Do Not Reject Null Hypothesis
STEP 7: Interpretation   *Summary*   There is not a statistically significant mean difference between the sample and the population (z, (n = 25) = 1.66, p < 0.01).

10. According to the APA, members with PhDs who hold full-time teaching positions have an average salary of $31,500 per year with a standard deviation of $3000. A union negotiator wishes to determine whether the mean salary for female members of the APA is lower than the population. Subsequently, a sample of 100 women finds that women holding teaching positions have a mean salary of $30,300. Is this significantly lower than the overall mean salary? Test the null hypothesis at a significance level of 0.05.

CHAPTER 5

ANSWER FOR QUESTION 10

Seven Steps for Hypothesis Testing

STEP 1: z-test
STEP 2: Hypotheses H_0: (μ) $\geq$ 31,500 H_1: (μ) < 31,500
STEP 3: Alpha Level (α) Alpha level is 0.05.
STEP 4: Critical Value (C.V.) C.V. = –1.64
STEP 5: Statistical Test z = –4.00
STEP 6: Reject Null Hypothesis
STEP 7: Interpretation *Summary* There is a statistically significant mean difference between the sample and the population, where women members of the APA are paid less than the population (z, (n = 100) = –4.00, p < 0.05).

REFERENCES

Fisher, R. A. 1990. *Statistical methods, experimental design, and scientific inference*. Oxford: Oxford University Press.

Glass, G. V., and K. D. Hopkins. 2008. *Statistical methods in education and psychology*. Boston: Allyn and Bacon.

Critical Values for z-test

Alpha Level	One-Tail	Two-Tail
α = 0.05	C.V. = +/–1.64	C.V. = 1.96/–1.96
α = 0.01	C.V. = +/–2.33	C.V. = 2.58/–2.58
α = 0.001	C.V. = +/–3.10	C.V. = 3.30/–3.30

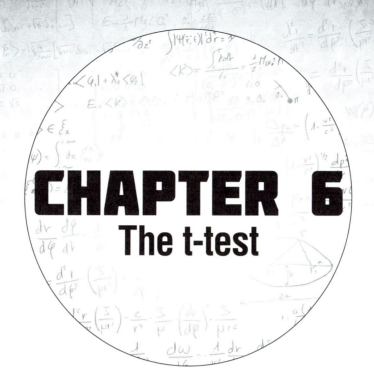

CHAPTER 6
The t-test

OBJECTIVES

When you complete this chapter, you should be able to
1. demonstrate an understanding of the one sample t-test, the dependent t-test, and the independent t-test to make inferences in classroom settings;
2. interpret t-tests and their relevance to classroom settings;
3. compare the relationship between descriptive statistics and parametric procedures;
4. calculate t-tests from classroom and standardized test results; and
5. find, use, and interpret t-test results.

Key Terms

When you complete this chapter, you should understand the following terms:
    degrees of freedom (df)   estimated standard error of the mean
    one sample t-test   one-tailed test
    dependent t-test   two-tailed test
    independent t-test

INTRODUCTION

If you were reading through old academic papers, you might see the original t-test referred to as "Student's t." Ironically, the t-test gets its name from William Sealy Gosset, an Oxford University alumnus in chemistry and mathematics, who worked for the Guinness Brewery in Dublin, Ireland. As detailed in chapter 1, Gosset developed the t-test to determine the difference between the independent samples. However, because Guinness had a policy protect their product, employees were not permitted to publish while employed. Subsequently, he published under the pseudonym of *A Student*.

Gosset took a leave of absence from Guinness and studied at the Galton Biomedical Laboratory under the famous statistician Karl Pearson. However, it was not long before Gosset made the acquaintance of Pearson's rival, Ronald Aylmer Fisher, which ignited progress in the field of statistics. Although it may be apocryphal, legend has it that the letter "t" in t-test has its origins in a summer tea party:

> At a summer tea party in Cambridge, England, a lady states that tea poured into milk tastes differently than that of milk poured into tea. Her notion is shouted down by the scientific minds of the group. But one guest, by the name Ronald Aylmer Fisher, proposes to scientifically test the lady's hypothesis. There was no better person to conduct such a test, for Fisher had brought to the field of statistics an emphasis on controlling the methods for obtaining data and the importance of interpretation. He knew that how the data was gathered and applied was as important as the data themselves (Salsburg 2001).

ONE-SAMPLE T-TEST

One of the most commonly used tests in hypothesis testing is the one-sample t-test. Like the z-test, the one-sample t-test assumes that the mean for a given population—such as "ITBS score" or "IQ"—is known. Once the population mean is known, the researcher can determine whether the population and sample means are significantly different. In this case, however, there is a sample distribution to extrapolate an estimated standard error using a sample standard deviation (See formula).

Formula:
SE = Estimated Standard Error

$$SE = \frac{s}{\sqrt{n}}$$

$$t = \frac{\bar{x} - \mu}{SE}$$

In this equation, $\bar{x}$ is the sample mean, μ is the population mean, s is the sample standard deviation (derived from sample data), and n is the size of the sample. With the one-sample t-test, the standard deviation of the sample is substituted for the standard deviation of the population. Therefore, the statistic does not have a normal distribution and the distribution is known as a t-distribution. Because there is a different t-distribution for each sample size, it is not practical to list a separate area-of-the-curve table for each one. Instead, critical t-values for common alpha levels (0.05, 0.01, 0.001, and so forth) are usually given in a single table for a range of sample sizes and can be used with one-tail and/or two-tail tests.

Values in the t-table are not actually listed by sample size but by ***degrees of freedom (df)***. The number of degrees of freedom for a problem involving the *t*-distribution for a one-sample t-test is calculated by taking the sample size (n) and subtracting 1 (n – 1). We will use the same seven steps that were used with hypothesis testing and the z-test.

Example 1

> **Example 1 (two-tailed test):** A ninth-grade science teacher wants to know how her students compare to other students in the state on the state exam. Ten students are chosen at random from the class and are given the state exam for general science. The state average on the test is a score of $\mu = 70$ on the test. The ten students get scores of 65, 93, 78, 69, 87, 77, 81, 85, 72, and 95. Please use the $\alpha = 0.05$.

To begin the calculation, we must follow a sequence of steps that will be used throughout hypothesis testing. As you will see, the most general way to compute a one-sample t-test is to define the statistical test statistic that can be calculated from scale of measurement and the collection of data. Below are the Seven Steps for Hypothesis Testing, which will be used in all of the statistical tests used throughout this book.

Seven Steps for Hypothesis Testing

STEP 1: Identify the appropriate statistical test to be used with the given data for analysis.

STEP 2: Hypotheses
H_o (Null Hypothesis): This hypothesis proposes that no statistical significance exists. The null hypothesis attempts to show that no variation exists between the means, or that a single variable is not different. When the null hypothesis is not rejected, it is presumed to be true until statistical evidence nullifies it for an alternative hypothesis.
H_1 (Alternative Hypothesis): The alternative hypothesis is the hypothesis that is contrary to the null hypothesis. This hypothesis should correspond with the researcher's assumption and is used to discern the conclusion.

STEP 3: Alpha Level (α)
Alpha level is the probability of making a type-I error (rejecting the null hypothesis when the null hypothesis is true). The lower this value is, the less chance that the researcher has of committing a type-I error. Also, an experiment should use a sample size large enough (i.e., $n = 20$) so that there is adequate statistical power. As a general rule, statistical analysis in education uses alpha levels of 0.05 or sometimes 0.01.

STEP 4: Critical Value (C.V.)
A number that causes rejection of the null hypothesis if a given test statistic is this number or more, and acceptance of the null hypothesis if the test statistic is smaller than this number. The critical value can be found by analyzing a statistical table related to the appropriate statistical test.

Continued on next page.

Continued from previous page.

STEP 5: Statistical Test
A statistical test provides a mechanism for making quantitative decisions. The intent is to determine whether there is enough evidence to "reject" the null hypothesis. Not rejecting may be a good result if we want to continue to act as if we "believe" the null hypothesis is true. Or it may be a disappointing result, possibly indicating we may not yet have enough data to "prove" something by rejecting the null hypothesis.

STEP 6: Reject or Do Not Reject the Null Hypothesis
By comparing the critical value to the calculated statistics, if the statistic exceeds the critical value, the null hypothesis is rejected.

STEP 7: Interpretation
Rejecting the Null Hypothesis may disclose that a independent variable/treatment did affect the dependent variable. However, not rejecting the null hypothesis may be a good result if we are looking for a result where no difference is desirable. Or it may be a disappointing result, possibly indicating we may not yet have enough data to "prove" something by rejecting the null hypothesis.

After understanding that the data must be analyzed by using a one sample t-test, we can proceed with the hypothesis.

STEP 1: Identify the Statistical Test
One-sample t-test

Most researchers test the null hypothesis (H_0), which is a statement that there is "no difference" between the sample mean and the population mean. Conversely, the alternative hypothesis (H_1) indicates that the sample mean and the population mean are not equal. In this case the hypotheses are written.

STEP 2: Hypotheses
$H_0: (\mu_{science\ scores}) = 70$
$H_1: (\mu_{science\ scores}) \neq 70$

The third step is to set the alpha level (α). As explained earlier, the alpha level is used to determine the probability of committing a type-I error. Therefore, different studies have advantages or disadvantages when using higher or lower α-levels. Hence, you have to be careful about interpreting the meaning of these terms. For example, *higher* α-levels can *increase* the chance of a type-I error. Therefore, a lower α-level indicates that you are conducting a test that is more rigorous. Keep in mind that an alpha level of 0.05 means that the researcher is taking a risk of being wrong 5 in 100 times by rejecting the null hypothesis (e.g., Committing a type-I error is saying there's an effect when there really is not). However, an alpha level (α) of 0.01 (compared with 0.05) means the researcher is being relatively cautious by minimizing the risk being wrong only 1 in 100 times to reject the null hypothesis. Based on the risk, the researcher decides the alpha level. In this case, the alpha level is ιο set at 0.05.

> **STEP 3: Alpha Level**
> α = 0.05

The fourth step is to determine the critical value. The t Distribution is used for the one sample t-test and is used to establish the critical value. First, you need to determine the direction of the statistical test or whether to use a one-tail or a two-tailed test. In this case, since no direction has been specified, the two-tailed test is used.

The t Distribution

Table entries are values of t corresponding to proportions in one-tail or in two-tails combined.

One-Tail	0.25	0.10	0.05	0.025	0.01	0.005
Two-Tail	0.50	0.20	0.10	0.05	0.02	0.01
df = 1	1.000	3.078	6.314	12.71	31.82	63.66
2	0.816	1.886	2.920	4.303	6.965	9.925
3	0.765	1.638	2.353	3.182	4.541	5.841
4	0.741	1.533	2.132	2.776	3.747	4.604
5	0.727	1.476	2.015	2.571	3.365	4.032
6	0.718	1.440	1.943	2.447	3.143	3.707
7	0.711	1.415	1.895	2.365	2.998	3.499
8	0.706	1.397	1.860	2.306	2.896	3.355
9	0.703	1.383	1.833	2.262	2.821	3.250

Since this is a two-tailed test and the sample size is 10, the degrees of freedom (df) equal 9 (i.e., df = 10 − 1 = 9). Once we have determined the degrees of freedom, we find the critical value by intersecting the degrees of freedom with the probability value from the tails column. Since this is a two-tailed test, we intersect the two-tail column for 0.05 and the df = 9, which gives a critical value of +/−2.262.

> **STEP 4: Critical Value**
> d.f. = 9
> C.V. = +/−2.262

The fifth step is to compute the t-test statistic. The formula for the t-test is:

SE = Estimated Standard Error = $\frac{s}{\sqrt{n}}$

$t = \frac{\bar{x} - \mu}{SE}$

In this equation, SE represents the Standard Error and estimates the amount of sampling error between the sample and the population. Since the state average on the test is a score of $\mu = 70$ on the test, we need to calculate a standard deviation from the ten students' scores of 65, 93, 78, 69, 87, 77, 81, 85, 72, and 95.

> **STEP 5: Statistical Test (t-test)**
> Step #1
> 65, 93, 78, 69, 87, 77, 81, 85, 72, 95
> $\bar{x}$ = 80.20
> S = 9.95
>
> SE = $\frac{9.95}{\sqrt{10}}$ = 3.15
>
> Step #2
> Next we calculate the one-sample t-test, which is the distance from the sample mean to the population mean in units of the standard error:
>
> $t = \frac{80.2 - 70}{3.15} = 3.24$

The calculation for the one-sample t-test = 3.24, which measures the difference between the science test scores of students at a local school and the science test scores of students from the population.

The sixth step is to reject or to not reject the null hypothesis. This step takes into consideration a comparison of the critical value to the calculated z-test statistic, which compares the one sample t-test of 3.24 to the C.V. of +/−2.262. Subsequently, since the t-test statistic exceeds the critical value of +/−2.262, the null hypothesis is rejected.

STEP 6: Reject or Do Not Reject the Null Hypothesis
Since the one sample t-test of 3.24 is larger than C.V. of +/− 2.262, we reject the null hypothesis.

The seventh step is to interpret our findings. Since we rejected our null hypothesis, we conclude that the observed sample mean is significantly different from the population. Therefore, we conclude that the science test scores of students enrolled in a local school are significantly higher than the science test scores of students from the population as a whole.

STEP 7: Interpretation
Rejecting the null hypothesis may disclose that an independent variable/treatment did affect the dependent variable.
There is a significant difference between the science test scores of students enrolled at a local school and the science test scores of students from the population.

Example 2

Example 2 (one-tailed test): A school district wants to know if an SAT prep program improves SAT scores for students in the school district. After participating in the program, twelve students are chosen at random and are given the SAT. The mean score for SAT is μ =1020 on the test. The 12 students get scores of 940, 1080, 1120, 1260, 1300, 720, 1020, 1150, 1050, 860, 1040, and 1410. Please use $\alpha = 0.05$.

STEP 1: Identify the Statistical Test
One-sample t-test

Most researchers test the null hypothesis (H_0), which is a statement that there is "no difference" between the sample mean and the population mean. However, in this case, since it is a one tailed test, if the mean is lower, this still reflects the null hypothesis. Conversely, the alternative hypothesis (H_1) indicates that the sample mean and the population mean are not equal and that the sample is greater than the population. In this case the hypotheses are written.

STEP 2: Hypotheses
H_0: (μ_{SAT}) ≤ 1020
H_1: (μ_{SAT}) > 1020

CHAPTER 6

The third step is to set the alpha level (α). As explained earlier, the alpha level is used to determine the probability of committing a Type 1 error. Therefore, different studies have advantages or disadvantages when using higher or lower α-levels. Based on the risk, the researcher decides the alpha level α. In this case, the alpha level α would be set at 0.05.

> **STEP 3: Alpha Level**
> α = 0.05

The fourth step is to determine the critical value. Since this is a one-tailed test (i.e., the research specified that we will look for a critical value for a t-test where students "have higher" SAT scores), we will use a one-tail test with an alpha level of .05.

The t Distribution

Table entries are values of t corresponding to proportions in one-tail or in two-tails combined.

One-Tail	0.25	0.10	0.05	0.025	0.01	0.005
Two-Tail	0.50	0.20	0.10	0.05	0.02	0.01
df = 1	1.000	3.078	6.314	12.71	31.82	63.66
2	0.816	1.886	2.920	4.303	6.965	9.925
3	0.765	1.638	2.353	3.182	4.541	5.841
4	0.741	1.533	2.132	2.776	3.747	4.604
5	0.727	1.476	2.015	2.571	3.365	4.032
6	0.718	1.440	1.943	2.447	3.143	3.707
7	0.711	1.415	1.895	2.365	2.998	3.499
8	0.706	1.397	1.860	2.306	2.896	3.355
9	0.703	1.383	1.833	2.262	2.821	3.250
10	0.700	1.372	1.812	2.228	2.764	3.169
11	0.697	1.363	1.796	2.201	2.718	3.106
12	0.695	1.356	1.782	2.179	2.681	3.055

For an alpha level of 0.05, the critical value = +1.796.

> **STEP 4: Critical Value**
> df = 11
> C.V. = +1.796

The fifth step is to compute the t-test statistic. The formula for the t-test is:

SE = Estimated Standard Error = $\frac{s}{\sqrt{n}}$

$t = \frac{\bar{x} - \mu}{SE}$

In this equation, SE represents the Standard Error and estimates the amount of sampling error between the sample and the population. The mean score for SAT is $\mu = 1020$ on the test. The ten students get scores of 940, 1080, 1120, 1260, 1300, 720, 1020, 1150, 1050, 860, 1040, and 1410.

> **STEP 5: Statistical Test (t-test)**
> Step #1
> 940, 1080, 1120, 1260, 1300, 720, 1020, 1150, 1050, 860, 1040, and 1410
> $\bar{x} = 1079.17$
> n = 12
> S = 190.67
>
> $SE = \frac{190.67}{\sqrt{12}} = 55.04$
>
> Step #2
> Next we calculate the one-sample t-test, which is the distance from the sample mean to the population mean in units of the standard error:
>
> $t = \frac{1079.17 - 1020}{55.04} = +1.08$

Again, since we used the same data, the calculation for the t-test = +1.08, which measures the SAT scores of students at a local school and the SAT scores of students within the population. The sixth step is to reject or to not reject the null hypothesis. This step takes into consideration a comparison of the critical value to the calculated t-test statistic, which compares the t-test of +1.08 to the C.V. of +1.796. Subsequently, since the t-test statistic does not exceed the critical value of +1.796, the null hypothesis is not rejected.

> **STEP 6: Reject or Do Not Reject the Null Hypothesis**
> Since the t-test of +1.08 does not exceed the C.V. of +1.796, we do not reject the null hypothesis.

The seventh step is to interpret our findings. Since we do not reject our null hypothesis, we conclude that the observed sample mean is not significantly different from the population. Therefore, we conclude that there is no difference between the SAT scores of local high school students and the SAT scores of the population.

> **STEP 7: Interpretation**
> Rejecting the null hypothesis may disclose that a independent variable/treatment did affect the dependent variable.
> There is no significant difference between the SAT scores of students at a local school district as compared to those within the population.

CALCULATOR EXPLORATION

Using the TI-83 or TI-84, you can easily compute a one-sample t-test. Using the one-tailed test from Example #2, we will demonstrate how to calculate a one-sample t-test with a graphing calculator.

Press [**STAT**].
Select [**TESTS**] [**1:T-Test…**]
Press [**ENTER**].

```
EDIT CALC TESTS
1:Z-Test…
2:T-Test…
3:2-SampZTest…
4:2-SampTTest…
5:1-PropZTest…
6:2-PropZTest…
7↓ZInterval…
```

```
T-Test
 Inpt:Data Stats
 μ0:1020
 List:L1
 Freq:1
 μ:≠μ0 <μ0 >μ0
 Calculate Draw
```

Highlight [**Data**] and enter the population mean, L1, and sample mean. Also, make sure that the greater than symbol is highlighted for a one tail test and press [**Calculate**].

Results

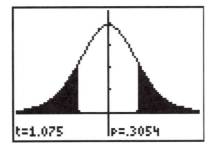

Graph with normal curve area is available if you select [**Draw**].

The t-test test statistic is +1.07. Since +1.07 < +1.796... and p-value is less than 0.05, we do not reject the null hypothesis.

INDEPENDENT T-TEST

Up to this point, all of the inferential statistical tests that we have examined have used a single sample to make inferences about a population. The independent t-test is designed to determine if there is a significant difference between two sample means (groups). Hence, several assumptions are possible:

- The subjects are assigned randomly into two independent groups;
- The subjects come from two completely separate groups;
- One of the groups is manipulated experimentally.

CHAPTER 6

Once the sample data are available, the researcher can determine whether sample means are significantly different and generalize the findings to the populations. In this case, there are two distinct sample distributions to extrapolate data (See formula).

Formulas:

$$\text{Independent-Measures } t = \frac{\text{Sample Mean Difference}}{\text{Standard Error for the Sample Mean Difference}}$$

or

$$t = \frac{\bar{x}_1 - \bar{x}_2}{S_{\bar{x}_1 - \bar{x}_2}}$$

$$S_{\bar{x}_1 - \bar{x}_2} = \sqrt{\frac{Sp^2}{df_1} + \frac{Sp^2}{df_2}}$$

$$\text{Pooled Variances} = Sp^2 = \frac{SS_1 + SS_2}{df_1 + df_2}$$

Degrees of Freedom:
$df = (n_1 + n_2) - 2$

In this equation, $\bar{x}_1$ represents the sample mean one and $\bar{x}_2$ represents the sample mean two. Moreover, **Sp^2** is the pooled variance and is based on a ratio of the combined sum of the squares (**SS**) from each and the collective degrees of freedom from the samples. One these are calculated, the **$S\bar{x}_1 - \bar{x}_2$** (Standard Error for the Sample Mean Difference) can be calculated. Finally, the degrees of freedom (df) is based on the equation $df = (n_1 + n_2) - 2$, which includes both groups.

The independent t-test is a way of determining the difference between two separate samples. For example, sometimes there are two distinct group (e.g., men and women or placebo group versus experimental group). Therefore, the independent t-test is used to determine whether a real difference exists between the groups and if not, if it is a result of sampling error. Hence, the hypothesis test offers a standardized and formal procedure for making this decision.

Example 1 (one-tailed independent t-test): An experiment is conducted to determine whether technology-based inquiry is more effective than traditional lectures. Two randomly chosen groups are instructed separately and then administered proficiency tests. Please use α = 0.05.

Group 1: Technology-Based Inquiry Group (n=10)	Group 2: Traditional Lecture Group (n = 10)
40	38
45	30
37	25
32	19
28	22
40	26
35	32
29	18
33	20
36	26

STEP 1: Identify the Statistical Test
Independent t-test

Most researchers test the null hypothesis (H_o), which is a statement that there is "no difference" between the sample mean and the population mean. However, in this case, since it is a one-tailed test, if the mean is lower, this still reflects the null hypothesis. Conversely, the alternative hypothesis (H_1) indicates that the technology-based inquiry group has a significantly higher mean than the traditional lecture group. Subsequently, if the null hypothesis is rejected, we can generalize that the difference exists in the population. In this case the hypotheses are written.

STEP 2: Hypotheses
$H_o: \mu_1 \leq \mu_2$
$H_1: \mu_1 > \mu_2$

The third step is to set the alpha level (α). As explained earlier, the alpha level is used to determine the probability of committing a type-I error. Therefore, different studies have advantages or disadvantages when using higher or lower α-levels. Based on the risk, the researcher decides the alpha level. In this case, the alpha level is set at 0.05.

CHAPTER 6

STEP 3: Alpha Level
$\alpha = 0.05$

The fourth step is to determine the critical value. Since this is a one-tailed test and the equation to calculate degrees of freedom (df) is:

$$df = (n1 + n2) - 2$$

Therefore, the degrees of freedom are:

$$df = (10 + 10) - 2 = 18$$

Again alpha level of 0.05. The critical value (C.V.) is 1.734.

The t Distribution

Table entries are values of t corresponding to proportions in one-tail or in two-tails combined

One-Tail	0.25	0.10	0.05	0.025	0.01	0.005
Two-Tail	0.50	0.20	0.10	0.05	0.02	0.01
df = 1	1.000	3.078	6.314	12.71	31.82	63.66
2	0.816	1.886	2.920	4.303	6.965	9.925
3	0.765	1.638	2.353	3.182	4.541	5.841
4	0.741	1.533	2.132	2.776	3.747	4.604
5	0.727	1.476	2.015	2.571	3.365	4.032
6	0.718	1.440	1.943	2.447	3.143	3.707
7	0.711	1.415	1.895	2.365	2.998	3.499
8	0.706	1.397	1.860	2.306	2.896	3.355
9	0.703	1.383	1.833	2.262	2.821	3.250
10	0.700	1.372	1.812	2.228	2.764	3.169
11	0.697	1.363	1.796	2.201	2.718	3.106
12	0.695	1.356	1.782	2.179	2.681	3.055

Continued on next page.

Continued from previous page.

13	0.694	1.350	1.771	2.160	2.650	3.012
14	0.692	1.345	1.761	2.145	2.624	2.977
15	0.691	1.341	1.753	2.131	2.602	2.947
16	0.690	1.337	1.746	2.120	2.583	2.921
17	0.689	1.333	1.740	2.110	2.567	2.898
18	0.688	1.330	1.734	2.101	2.552	2.878

Group 1: Technology-Based Inquiry Group (n = 10)	Group 2: Traditional Lecture Group (n = 10)
40	38
45	30
37	25
32	19
28	22
40	26
35	32
29	18
33	20
36	26

For an alpha level of 0.05, the critical value = +1.796.

> **STEP 4: Critical Value**
> df = 18
> C.V. = +1.734

The fifth step is to compute the t-test statistic. The calculations are based on the following data:

STEP 5: Statistical Test (Independent t-test)

Step 1
Compute the means:
$\bar{x}1 =$
$\bar{x}2 =$

Step 2
Pooled Variances = $Sp^2 = \dfrac{SS1 + SS2}{df1 + df2}$

Step 3

$S\bar{x}1 - \bar{x}2 = \sqrt{\dfrac{Sp^2}{n1} + \dfrac{Sp^2}{n2}}$

Step 4

$t = \dfrac{\bar{x}1 - \bar{x}2}{S\bar{x}1 - \bar{x}2}$

Calculation Step 1
$\bar{x}1 = 35.50$
$\bar{x}2 = 25.60$

Calculation Step 2
Pooled Variances = $Sp^2 = \dfrac{250.56 + 360.39}{9 + 9} = 33.94$

Calculation Step 3

$S\bar{x}1 - \bar{x}2 = \sqrt{\dfrac{33.94}{10} + \dfrac{33.94}{10}} = 2.60$

Calculation Step 4

$t = \dfrac{35.50 - 25.60}{2.60} = +3.80$

In this equation, [$S\bar{x}1 - \bar{x}2$] represents the Estimated Standard Error and estimates the amount of sampling error between the sample and the population. To calculate the Estimated Standard Error, we must first calculate the Pooled Variance Sp^2.

The calculation for the t-test = +3.80, which measures how much the scores for the technology-based inquiry group exceeded the students in the tradition lecture group. The sixth step is to reject or to not reject the null hypothesis. This step takes into consideration a comparison of the critical value to the calculated independent t-test statistic, which compares the t-test of +3.80 to the C.V. of +1.734. Subsequently, since the t-test statistic exceeds the critical value of +1.734, the null hypothesis is rejected.

> **STEP 6: Reject or Do Not Reject the Null Hypothesis**
> Since the t-test of +3.80 exceeds the C.V. of +1.734, we reject the null hypothesis.

The seventh step is to interpret our findings. Since reject our null hypothesis, we conclude that the test scores for students in the technology-based inquiry group are significantly higher than those students in the tradition lecture group. Therefore, we conclude that there is a significant difference between the technology-based inquiry group the tradition lecture group, where the technology-based inquiry group had higher scores.

> **STEP 7: Interpretation**
> Rejecting the null hypothesis may disclose that an independent variable/treatment did affect the dependent variable.
> The test scores for students in the technology-based inquiry group are significantly higher than those students in the tradition lecture group.

> **Example 2 (two-tailed independent t-test):** A school psychologist wants to find out if there is a difference between the scholastic achievement levels of students who participate in preschool programs versus students who do not participate in those programs. Please use $\alpha = 0.05$.

Group 1: Preschool Group ($n = 12$)	Group 2: Nonpreschool Group ($n = 12$)
8	8
9	6
6	7
9	7
6	5
7	8
9	5
7	5
8	7
7	6
8	6
8	6

STEP 1: Identify the Statistical Test
Independent t-test

Most researchers test the null hypothesis (H_o), which is a statement that there is "no difference" between the sample mean and the population mean. However, in this case, since it is a two-tailed test, if the mean is lower, this still reflects the null hypothesis. Conversely, the alternative hypothesis (H_1) indicates that the preschool group has a significantly higher mean than the nonpreschool group. Subsequently, if the null hypothesis is rejected, we can generalize that the difference exists in the population. In this case the hypotheses are written.

STEP 2: Hypotheses
H_o: $\mu_1 = \mu_2$
H_1: $\mu_1 \neq \mu_2$

The third step is to set the alpha level (α). As explained earlier, the alpha level is used to determine the probability of committing a type-I error. Therefore, different studies have advantages or disadvantages when using higher or lower α-levels. Based on the risk, the researcher decides the alpha level. In this case, the alpha level is set at 0.05.

STEP 3: Alpha Level
$\alpha = 0.05$

The fourth step is to determine the critical value. Since this is a one-tailed test and the equation to calculate degees of freedom (df) is:
$$df = n1 + n2 - 2$$

Therefore, the degrees of freedom are:
$$df = 12 + 12 - 2 = 22$$

The t Distribution

Table entries are values of t corresponding to proportions in one-tail or in two-tails combined

One-Tail	0.25	0.10	0.05	0.025	0.01	0.005
Two-Tail	0.50	0.20	0.10	0.05	0.02	0.01
df = 1	1.000	3.078	6.314	12.71	31.82	63.66
2	0.816	1.886	2.920	4.303	6.965	9.925
3	0.765	1.638	2.353	3.182	4.541	5.841
4	0.741	1.533	2.132	2.776	3.747	4.604
5	0.727	1.476	2.015	2.571	3.365	4.032
6	0.718	1.440	1.943	2.447	3.143	3.707
7	0.711	1.415	1.895	2.365	2.998	3.499
8	0.706	1.397	1.860	2.306	2.896	3.355
9	0.703	1.383	1.833	2.262	2.821	3.250
10	0.700	1.372	1.812	2.228	2.764	3.169
11	0.697	1.363	1.796	2.201	2.718	3.106
12	0.695	1.356	1.782	2.179	2.681	3.055
13	0.694	1.350	1.771	2.160	2.650	3.012
14	0.692	1.345	1.761	2.145	2.624	2.977
15	0.691	1.341	1.753	2.131	2.602	2.947
16	0.690	1.337	1.746	2.120	2.583	2.921
17	0.689	1.333	1.740	2.110	2.567	2.898
18	0.688	1.330	1.734	2.101	2.552	2.878
19	0.688	1.328	1.729	2.093	2.539	2.861
20	0.687	1.325	1.725	2.086	2.528	2.845
21	0.686	1.323	1.721	2.080	2.518	2.831
22	0.686	1.321	1.717	2.074	2.508	2.819

Again the alpha level is set at 0.05 so the critical value (C.V.) is 2.074 (or +/− 2.074). When we refer to the t distribution table and intersect the df = 22 for the two-tailed test, we find that the critical value = +/−2.074.

STEP 4: Critical Value
df = 22
C.V. = +/−2.074

The fifth step is to compute the t-test statistic. The calculations are based on the following data:

STEP 5: Statistical Test (Independent t-test)

Step 1
Compute the means:
$\bar{x}_1 =$
$\bar{x}_2 =$

Step 2
Pooled Variances = $Sp^2 = \dfrac{SS1 + SS2}{df1 + df2}$

Step 3
$\bar{x}_1 - \bar{x}_2 = \sqrt{\dfrac{Sp^2}{n1} + \dfrac{Sp^2}{n2}}$

Step 4
$t = \dfrac{\bar{x}_1 - \bar{x}_2}{S\bar{x}_1 - \bar{x}_2}$

Calculation Step 1
$\bar{x}_1 = 7.67$
$\bar{x}_2 = 6.25$

Calculation Step 2
Pooled Variances = $Sp^2 = \dfrac{12.67 + 14.25}{11 + 11} = 1.22$

Calculation Step 3
$\bar{x}_1 - \bar{x}_2 = \sqrt{\dfrac{1.22}{12} + \dfrac{1.22}{12}} = .451$

Calculation Step 4
$t = \dfrac{7.67 - 6.25}{.451} = +3.15$

In this equation, $S_{\bar{x}1-\bar{x}2}$ represents the Estimated Standard Error and estimates the amount of sampling error between the sample and the population. To calculate the Estimated Standard Error, we must first calculate the Pooled Variance Sp^2.

The calculation for the t-test = +3.15, which shows that the scores for the preschool group exceeded the students in the non-preschool group. The sixth step is to reject or do not reject the null hypothesis. This step takes into consideration a comparison of the critical value to the calculated independent t-test statistic, which compares the t-test of +3.15 to the C.V. of +/–2.074. Subsequently, since the t-test statistic exceeds the critical value of +/–2.074, the null hypothesis is rejected.

> **STEP 6: Reject or Do Not Reject the Null Hypothesis**
> Since the t-test of +3.15 exceeds the C.V. of +/– 2.074, we reject the null hypothesis.

The seventh step is to interpret our findings. Since we reject our null hypothesis, we conclude that the test scores for students in the preschool group are significantly higher than those students in the non-preschool group. Therefore, we conclude that there is a significant difference between the preschool group the nonpreschool group, where the students in the preschool group had higher scores.

> **STEP 7: Interpretation**
> Rejecting the null hypothesis may disclose that a independent variable/treatment did affect the dependent variable.
> The test scores for students in the preschool group are significantly higher than those students in the nonpreschool group.

CALCULATOR EXPLORATION

Using the TI-83 or TI-84, you can easily compute an independent t-test. Using the two-tailed test from Example 1, we will demonstrate how to calculate a t-test with a graphing calculator.

Press [**STAT**].
Select [**EDIT**].
Press [**ENTER**].

L1	L2	L3
40	38	------
45	30	
37	25	
32	19	
28	22	
40	26	
35	32	

L1 ={40,45,37,32...

Press [**STAT**].
Select [**TESTS 4:2-SampTTest…**].
Press [**ENTER**].

Highlight [**Data**] and enter the L1, and Freq L1. Also, make sure that the greater than symbol is highlighted for a one tail test and Press [**Calculate**].

Results

Graph with normal curve area is available if you select [**Draw**].

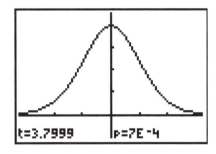

The t-test test statistic is +3.799. Since +3.799 > +1.734, and p-value is less than 0.05, we reject the null hypothesis.

Dependent t-Test

The dependent t-test is commonly used when two scores are recorded from the same individuals. For example, weight measurements might be taken for each subject at the beginning (pre-test) and again at the end of a program (e.g., weight lifting to determine if there has been a weight change). Sometimes the dependent t-test is referred to as the related measures t-test or the matched pairs t-test.

> **Example 1 (two-tailed dependent t-test):** A scientist conducts an experiment that seeks to find out if an exercise program has an effect on the time it takes for people to walk up a flight of stairs (time in minutes). Please use $\alpha = 0.05$.

Subject	Measurement 1: Before the Exercise Program ($n = 11$)	Measurement 2: After the Exercise Program ($n = 11$)
Bob	3.25	3.00
Merv	2.14	2.15
Susan	2.87	2.53
Kay	1.32	1.35
Becky	0.41	0.00
John	1.59	1.44
Lou	1.33	1.30
George	3.01	2.50
Alex	2.87	0.43
Ron	2.53	2.21
Pam	1.77	0.89

> **STEP 1: Identify the Statistical Test**
> Dependent t-test

Most researchers test the null hypothesis (H_o), which is a statement that there is "no difference" between the sample mean and the population mean. However, in this case, since it is a two-tailed test, if the mean is lower, this still reflects the null hypothesis. Conversely, the alternative hypothesis (H_1) indicates that the first measurement has a significantly different mean than the second measurement. Subsequently, if the null hypothesis is rejected, we can generalize that the difference exists in the population. In this case the hypotheses are written.

> **STEP 2: Hypotheses**
> $H_o: \mu_D = 0$
> $H_1: \mu_D \neq 0$

The third step is to set the alpha level (α). As explained earlier, the alpha level is used to determine the probability of committing a type-I error. Therefore, different studies have advantages or disadvantages when using higher or lower α-levels. Based on the risk, the researcher decides the alpha level α. In this case, the alpha level is set at 0.05.

> **STEP 3: Alpha Level**
> $\alpha = 0.05$

The fourth step is to determine the critical value. Since this is a two-tailed test and the equation to calculate degrees of freedom (df) is:

$$df = n - 1$$

Therefore, the degrees of freedom are:

$$df = 11 - 1 = 10$$

The t Distribution

Table entries are values of t corresponding to proportions in one-tail or in two-tails combined.

One-Tail	0.25	0.10	0.05	0.025	0.01	0.005
Two-Tail	0.50	0.20	0.10	0.05	0.02	0.01
df = 1	1.000	3.078	6.314	12.71	31.82	63.66
2	0.816	1.886	2.920	4.303	6.965	9.925
3	0.765	1.638	2.353	3.182	4.541	5.841
4	0.741	1.533	2.132	2.776	3.747	4.604
5	0.727	1.476	2.015	2.571	3.365	4.032
6	0.718	1.440	1.943	2.447	3.143	3.707
7	0.711	1.415	1.895	2.365	2.998	3.499
8	0.706	1.397	1.860	2.306	2.896	3.355
9	0.703	1.383	1.833	2.262	2.821	3.250
10	0.700	1.372	1.812	2.228	2.764	3.169

Again, alpha level of 0.05. The critical value (C.V.) is 2.228 (or +/−2.228), we refer to the t distribution table and intersect the df = 10.

STEP 4: Critical Value
df = 10
C.V. = +/−2.228

The fifth step is to compute the t-test statistic. The calculations are based on the following data:

STEP 5: Statistical Test (Independent t-test)

Equations:

$$t = \frac{\overline{D}}{s_{\overline{D}}}$$

$$s_{\overline{D}} = \frac{s_D}{\sqrt{n}}$$

$s_{\overline{D}}$ is the standard error of the difference scores.

Step 1

Compute the difference (D):

3.25 – 3.00 = 0.25
2.14 – 2.15 = -0.01
2.87 – 2.53 = 0.34
1.32 – 1.35 = -0.03
0.41 – 0.00 = 0.41
1.59 – 1.44 = 0.15
1.33 – 1.30 = 0.03
3.01 – 2.50 = 0.51
2.87 – 2.43 = 0.44
2.53 – 2.21 = 0.32
1.77 – 0.89 = 0.88

Step 2

$$s_{\overline{D}} = \frac{s_D}{\sqrt{n}}$$

$$s_{\overline{D}} = \frac{2.68}{\sqrt{11}} = 0.081$$

Step 3

$$\overline{D} = \text{Average difference} = \frac{\sum D}{n} = \frac{3.29}{11} = 0.299$$

Step 4

$$t = \frac{\overline{D}}{s_{\overline{D}}} = \frac{0.299}{0.081} = +\,3.69$$

In this equation, [$S\overline{D}$] represents the Estimated Standard Error and estimates the amount of sampling error between the sample and the population. To calculate the Estimated Standard Error, we must first calculate the Difference (D), which is based on subtracting the differences between measurement 1 and measurement 2.

The calculation for the t-test = +3.69, which shows that the scores for measurement 1 exceeded measurement 2. The sixth step is to reject or do not reject the null hypothesis. This step takes into consideration a comparison of the critical value to the calculated independent t-test statistic, which compares the t-test of +3.69 to the C.V. of +/−2.228. Subsequently, since the t-test statistic exceeds the critical value of +/−2.074, the null hypothesis is rejected.

> **STEP 6: Reject or Do Not Reject the Null Hypothesis**
> Since the t-test of +3.69 exceeds the C.V. of +/− 2.228, we reject the null hypothesis.

The seventh step is to interpret our findings. Since reject our null hypothesis, we conclude that the test scores for measurement 1 are significantly higher than those for measurement 2. Therefore, we conclude that there is a significant difference between the groups, where the measurement 1 had higher scores.

> **STEP 7: Interpretation**
> Rejecting the null hypothesis may disclose that a independent variable/treatment did affect the dependent variable.
> The scores for measurement 1 are significantly higher than measurement 2. Therefore, without an exercise program, we can conclude that participants in the first group took more time to walk up a flight of stairs.

> **Example #2 (one-tailed dependent t-test):** A scientist conducts an experiment that seeks to find out if a weight loss program is successful (weight is in pounds). Please use $\alpha = 0.05$.

CHAPTER 6

Subject	Measurement 1: Before the Weight Loss Program ($n = 11$)	Measurement 2: After the Weight Loss Program ($n = 11$)
Fred	176	166
Joe	240	236
Alice	155	156
Ruth	183	146
Madeline	120	119
Steve	00	276
Tommy	190	181
Brent	230	223
Mary	165	160
Matt	285	255
Betty	200	190

STEP 1: Identify the Statistical Test
Dependent t-test

Most researchers test the null hypothesis (H_o), which is a statement that there is "no difference" between the sample mean and the population mean. However, in this case, since it is a one tailed test, if the mean is lower, this still reflects the null hypothesis. Conversely, the alternative hypothesis (H_1) indicates that the first measurement has a significantly different mean than the second measurement. Subsequently, if the null hypothesis is rejected, we can generalize that the difference exists in the population. In this case the hypotheses are written.

STEP 2: Hypotheses
$H_o: \mu_D \geq 0$
$H_1: \mu_D < 0$

The third step is to set the alpha level (α). As explained earlier, the alpha level is used to determine the probability of committing a type-I error. Therefore, different studies have advantages or disadvantages when using higher or lower α-levels. Based on the risk, the researcher decides the alpha level α. In this case, the alpha level is set at 0.05.

STEP 3: Alpha Level
$\alpha = 0.05$

The fourth step is to determine the critical value. Since this is a one-tailed test and the equation to calculate degees of freedom (df) is:

$$df = n - 1$$

Therefore, the degrees of freedom are:

$$df = 11 - 1 = 10$$

The t Distribution

Table entries are values of t corresponding to proportions in one-tail or in two-tails combined

One-Tail	0.25	0.10	0.05	0.025	0.01	0.005
Two-Tail	0.50	0.20	0.10	0.05	0.02	0.01
df = 1	1.000	3.078	6.314	12.71	31.82	63.66
2	0.816	1.886	2.920	4.303	6.965	9.925
3	0.765	1.638	2.353	3.182	4.541	5.841
4	0.741	1.533	2.132	2.776	3.747	4.604
5	0.727	1.476	2.015	2.571	3.365	4.032
6	0.718	1.440	1.943	2.447	3.143	3.707
7	0.711	1.415	1.895	2.365	2.998	3.499
8	0.706	1.397	1.860	2.306	2.896	3.355
9	0.703	1.383	1.833	2.262	2.821	3.250
10	0.700	1.372	1.812	2.228	2.764	3.169

Again, alpha level of 0.05. The critical value (C.V.) is -1.821, we refer to the t distribution table and intersect the df = 10.

> **STEP 4: Critical Value**
> df = 10
> C.V. = -1.821

The fifth step is to compute the t-test statistic. The calculations are based on the following data:

STEP 5: Statistical Test (Independent t-test)

Equations:

$$t = \frac{\overline{D}}{s_{\overline{D}}}$$

$$s_{\overline{D}} = \frac{s_D}{\sqrt{n}}$$

$s_{\overline{D}}$ is the standard error of the difference scores.

Step 1

Compute the difference (D):

176 – 166 = 10
240 – 236 = 4
155 – 156 = -1
183 – 146 = 37
120 – 119 = 1
300 – 276 = 24
190 – 181 = 9
230 – 223 = 7
165 – 160 = 5
285 – 255 = 30
200 – 190 = 10

Step 2

$$s_{\overline{D}} = \frac{s_D}{\sqrt{n}}$$

$$s_{\overline{D}} = \frac{12.39}{\sqrt{11}} = 3.74$$

Step 3

$$\overline{D} = \text{Average difference} = \frac{\sum \overline{D}}{n} = \frac{136}{11} = 12.36$$

Step 4

$$t = \frac{\overline{D}}{s_{\overline{D}}} =$$

$$t = \frac{\overline{D}}{s_{\overline{D}}} = \frac{12.36}{3.74} = 3.31$$

In this equation, [S $\overline{D}$] represents the Estimated Standard Error and estimates the amount of sampling error between the sample and the population. To calculate the Estimated Standard Error, we must first the Differences (D), which is based on subtracting the differences between measurement 1 and measurement 2. The calculation for the t-test = 3.31, which shows that the scores for measurement 1 exceeded measurement 2.

The sixth step is to reject or do not reject the null hypothesis. This step takes into consideration a comparison of the critical value to the calculated independent t-test statistic, which compares the t-test of 3.31 (the t-test is negative since the scores decreased. Therefore, it will be designated as –3.31) to the C.V. of –1.812. Subsequently, since the t-test statistic exceeds the critical value of –1.812, the null hypothesis is rejected.

> **STEP 6: Reject or Do Not Reject the Null Hypothesis**
> Since the t-test of –3.31 exceeds the C.V. of –1.812, we reject the null hypothesis.

The seventh step is to interpret our findings. Since reject our null hypothesis, we conclude that the weights for measurement 1 are significantly higher than those for measurement 2. Therefore, we conclude that there is a significant difference between the groups, where the weight loss program resulted in a significant weight loss for the participants.

> **STEP 7: Interpretation**
> Rejecting the null hypothesis may disclose that a independent variable/treatment did affect the dependent variable.
> The scores for measurement 2 are significantly lower than measurement 1. Therefore, the weight loss program resulted in a significant reduction of weight.

CALCULATOR EXPLORATION

Using the TI-83 or TI-84, you can easily compute a dependent t-test. Using the two-tailed test from Example #1, we will demonstrate how to calculate a z-test with a graphing calculator.

Press [**STAT**].
Select [**EDIT**].
Press [**ENTER**].

Press [**L3**].
At the bottom input [**L3 = L1-L2**].
Press [**ENTER**].

The result will be:

Press [**STAT**].
Press [**T-Test**].
Press [**Enter**].

Highlight [**DATA**].
Input [**L3**].

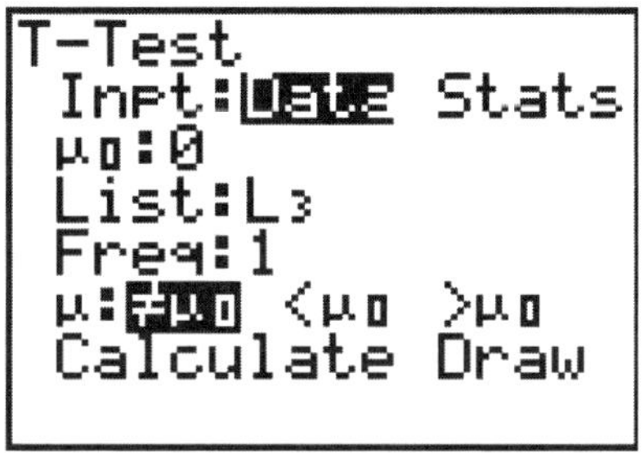

Press [**Calculate**].
  Results:

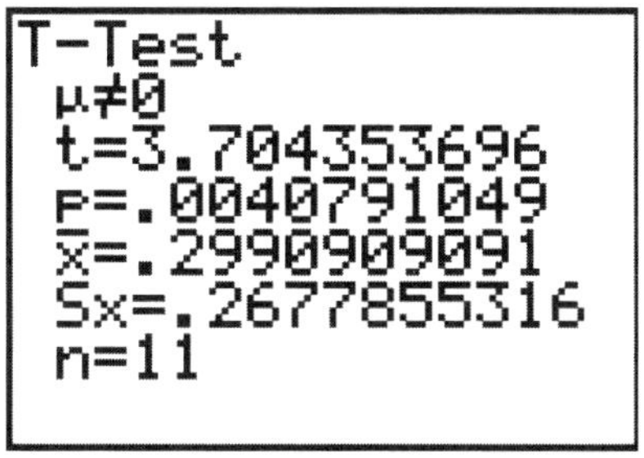

The t-test test statistic is +3.704. Since +3.704 > +/–2.228, and p-value is less than 0.05, we reject the null hypothesis.

SUMMARY

This chapter introduces the t-test, which is a statistical tool for hypothesis testing. A t-test for one sample is used to compare the mean of a sample to a population for which the mean is known. However, unlike the z-test, the population variance is estimated from the scores in the sample.

The independent t-test uses data from two separate groups and focuses on a research question that attempts to evaluate the possibility of a significant mean difference. However, the dependent t-test compares two sets of scores from a repeated-measures or within-subjects design (a research design in which you have two scores from each individual in your sample or research participants that are paired in some way).

Subsequently, the t-test is a tool for researchers to compare means and generalize findings to a population.

Seven Steps for Hypothesis Testing

STEP 1: Identify the appropriate statistical test to be used with the given data for analysis.
STEP 2: Hypotheses • **H_o (Null Hypothesis):** This hypothesis proposes that no statistical significance exists. The null hypothesis attempts to show that no variation exists between the means, or that a single variable is not different. When the null hypothesis is not rejected, it is presumed to be true until statistical evidence nullifies it for an alternative hypothesis. • **H_1 (Alternative Hypothesis):** The alternative hypothesis is the hypothesis that is contrary to the null hypothesis. This hypothesis should correspond with the researcher's assumption and is used to discern the conclusion.
STEP 3: Alpha Level (α) Alpha Level is the probability of making a type-I error (rejecting the null hypothesis when the null hypothesis is true). The lower that this value is, the less chance that the researcher has of committing a Type I error. Also, an experiment should use a sample size large enough (i.e., $n = 20$) so that there is adequate statistical power. As a general rule, statistical analysis in education uses alpha levels of 0.05 or sometimes 0.01.
STEP 4: Critical Value (C.V.) A number that causes rejection of the null hypothesis if a given test statistic is this number or more, and nonrejection of the null hypothesis if the test statistic is smaller than this number. The critical value can be found by analyzing a statistical table related to the appropriate statistical test.
STEP 5: Statistical Test A statistical test provides a mechanism for making quantitative decisions. The intent is to determine whether there is enough evidence to "reject" the null hypothesis. Not rejecting may be a good result if we want to continue to act as if we "believe" the null hypothesis is true. Or it may be a disappointing result, possibly indicating we may not yet have enough data to "prove" something by rejecting the null hypothesis.

> **STEP 6: Reject or Do Not Reject the Null Hypothesis**
> By comparing the critical value to the calculated statistics, if the statistic exceeds the critical value, the null hypothesis is rejected.
>
> **STEP 7: Interpretation**
> Rejecting the null hypothesis may disclose that an independent variable/treatment did affect the dependent variable. However, not rejecting the null hypothesis may be a good result if we are looking for a result where no difference is desirable. Or it may be a disappointing result, possibly indicating we may not yet have enough data to "prove" something by rejecting the null hypothesis.

The phrase "significance test" was first mentioned by Ronald Fisher, when he explored whether there is difference between two means. In hypothesis testing, according to a pre-determined upper limit probability, a result is called statistically significant if it is unlikely to have occurred by chance alone. Hence, hypothesis testing is a fundamental technique for statistical inference and helps researchers use data to optimize decisions by minimizing the risk of type-I error (rejecting the null hypothesis when the null hypothesis is true).

Keep in mind that when a researcher thinks that a treatment will affect scores in a certain direction (an increase or a decrease), the research may choose a directional hypothesis (one-tail rather than two-tail). Next, the research will determine the power of the hypothesis test (alpha level), which is the probability of committing a type-I error. After the alpha level is selected, the critical value is selected from the z-distribution table (e.g., a one-tailed test with an alpha level of 0.001 gives a critical value of +/–3.30). The next step is to calculate a t-test and compare the z-test to the critical value (C.V.). If the t-test exceeds the C.V., the null hypothesis is rejected and an interpretation follows.

CHAPTER REVIEW QUESTIONS

Use Seven Steps for Hypothesis Testing to solve the following problems.

1. Determine whether a textbook makes a difference for students to learn science. Students average $\mu = 70$ on a national science exam. At the alpha level of 0.05, the results from a random sample of students who use the new textbook with the following scores on the national exam:
63, 70, 91, 70, 95, 81, 71, 65, 99

ANSWER FOR QUESTION 1

Seven Steps for Hypothesis Testing

STEP 1: One-sample t-test
STEP 2: Hypotheses $H_0: (\mu_{science\ scores}) = 70$ $H_1: (\mu_{science\ scores}) \neq 70$
STEP 3: Alpha Level (α) Alpha level is 0.05.
STEP 4: Critical Value (C.V.) C.V. = +/−2.306
STEP 5: Statistical Test t = 1.839
STEP 6: Do Not Reject Null Hypothesis
STEP 7: Interpretation *Summary* There is no statistically significant mean difference between the sample and the population (t, (n = 9) = 1.839, p > 0.05).

2. A researcher predicts that students who are exposed to loud noises experience a decrease in their hearing. On the standard test for hearing, those tested earn a hearing score of $\mu = 20$. Determine whether or not the sample of students has statistically lower hearing score from the population at a significance level of 0.01. Below are the scores from those in the sample:
14, 15, 16, 13, 11, 19, 16, 17, 8, 15

ANSWER FOR QUESTION 2

Seven Steps for Hypothesis Testing

STEP 1: One-sample t-test
STEP 2: Hypotheses H_o: ($\mu_{hearing\ scores}$) = 20 H_1: ($\mu_{hearing}$) ≠ 20
STEP 3: Alpha Level (α)   Alpha level is 0.01
STEP 4: Critical Value (C.V.)   C.V. = –2.821
STEP 5: Statistical Test   t = –5.65
STEP 6: Reject Null Hypothesis
STEP 7: Interpretation   *Summary*   There is a statistically significant mean difference between the sample and the population (t, (n =10) = –5.65, p > 0.01).

3. A scientist predicts that birds that eat seed with higher levels of protein gain weight. Determine whether or not the sample of birds has statistically higher mass from the population at a significance level of 0.05. Here is the mass (kg) for the birds who consumed seed with higher levels of protein (μ= 1.5 kg): 1.2, 1.5, 1.6, 1.3, 1.1, 1.9, 1.6, 1.7, 0.8, 1.5

ANSWER FOR QUESTION 3

Seven Steps for Hypothesis Testing

STEP 1: one-sample t-test
STEP 2: Hypotheses H_o: $(\mu_{birds}) \leq 1.5$ H_1: $(\mu_{birds}) > 1.5$
STEP 3: Alpha Level (α) Alpha level is 0.05.
STEP 4: Critical Value (C.V.) C.V. = +1.833
STEP 5: Statistical Test t = 0.785
STEP 6: Do Not Reject Null Hypothesis
STEP 7: Interpretation *Summary* There is no statistically significant mean difference between the sample and the population (t, (n = 10) = 0.785, p > 0.05).

4. An educational researcher examines the difference between students who use inquiry-based learning and those who are exposed to traditional lecture in a science classroom. Determine whether or not there is a difference between the two samples at a significance level of 0.05. Here are the test scores for the groups:
   Inquiry-based group [65, 75, 77, 80, 74, 86, 99, 81, 91, 75]
   Traditional lecture group [75, 75, 57, 68, 74, 86, 89, 81, 81, 65]

ANSWER FOR QUESTION 4

Seven Steps for Hypothesis Testing

STEP 1: Independent t-test
STEP 2: Hypotheses $H_0: \mu_1 = \mu_2$ $H_1: \mu_1 \neq \mu_2$
STEP 3: Alpha Level (α)   Alpha level is 0.05.
STEP 4: Critical Value (C.V.)   C.V. = +/−2.101
STEP 5: Statistical Test   t = 1.196
STEP 6: Do Not Reject Null Hypothesis
STEP 7: Interpretation   *Summary*   There is no statistically significant mean difference between the sample inquiry-based learning group and the traditional lecture group (t, (df = 18) = 1.196, p > 0.05).

5. An research corporation explores the difference in energy levels between a group that gets eight hours of sleep versus a group that gets six hours of sleep. Determine whether or not there is a difference between the energy levels of the two samples at a significance level of 0.01. Here are the test scores for the groups:
Eight-hour group [6, 7, 7, 8, 7, 8, 9, 8, 9, 7, 6]
Six-hour group [4, 5, 5, 6, 4, 6, 8, 1, 2, 5, 3]

ANSWER FOR QUESTION 5

Seven Steps for Hypothesis Testing

STEP 1: Independent t-test
STEP 2: Hypotheses $H_o: \mu_1 = \mu_2$ $H_1: \mu_1 \neq \mu_2$
STEP 3: Alpha Level (α)   Alpha level is 0.01.
STEP 4: Critical Value (C.V.)   C.V. = +/−2.528
STEP 5: Statistical Test   t = 4.474
STEP 6: Reject Null Hypothesis
STEP 7: Interpretation   *Summary*   There is a statistically significant mean difference between the group that had eight hours of sleep and the group that had six hours of sleep. The group that had eight hours of sleep had significantly higher energy levels (t, (df = 20) = 4.474, p > 0.01).

6. A researcher examines the benefits of a school breakfast program and would like to determine if having breakfast improves the achievement scores of students. Determine whether or not there is a difference in achievement between the two samples at a significance level of 0.05. Here are the test scores for the groups:
Breakfast group [96, 77, 87, 85, 87, 78, 91, 78, 89, 87, 66, 77]
No Breakfast group [84, 75, 85, 76, 84, 68, 58, 71, 62, 65, 73, 63]

ANSWER FOR QUESTION 6

Seven Steps for Hypothesis Testing

STEP 1: Independent t-test
STEP 2: Hypotheses $H_o: \mu_1 \leq \mu_2$ $H_1: \mu_1 > \mu_2$
STEP 3: Alpha Level (α)   Alpha level is 0.05.
STEP 4: Critical Value (C.V.)   C.V. = +1.717
STEP 5: Statistical Test   t = 3.151
STEP 6: Reject Null Hypothesis
STEP 7: Interpretation   *Summary*   There is a statistically significant mean difference between the group that had breakfast and the group did not. The group that had breakfast had significantly higher achievement levels (t, (df = 22) = 3.151, p > 0.05).

7. A researcher examines the effectiveness of a new cholesterol medication on a group of subjects over a six month period. At the significance level of 0.01, determine whether the drug decreases the cholesterol level of the subjects. Here are the cholesterol levels of the group:
Pretest cholesterol levels [200, 250, 270, 230, 240, 188, 300, 270, 274, 330]
Posttest cholesterol levels [150, 190, 210, 175, 190, 144, 220, 200, 196, 280]

ANSWER FOR QUESTION 7

Seven Steps for Hypothesis Testing

STEP 1: Dependent t-test
STEP 2: Hypotheses H_o: $\mu_D \geq 0$ H_1: $\mu_D < 0$
STEP 3: Alpha Level (α) Alpha level is 0.01.
STEP 4: Critical Value (C.V.) C.V. = -2.821
STEP 5: Statistical Test t = 15.12
STEP 6: Reject Null Hypothesis
STEP 7: Interpretation **Summary** There is a statistically significant mean difference between the pretest and posttest cholesterol levels. The posttest levels are significantly lower (t, (df = 9) = 15.12, p > 0.01).

8. A fitness trainer determines whether a supplement decreases the 40-meter dash times of a group of athletes. At the significance level of 0.05, determine whether the supplement decreases the times of the subjects. Here are the 40-meter dash times of the subjects of the group:

   Pretest times [4.9, 5.1, 5.2, 5.0, 5.5, 4.4., 5.8, 5.5, 4.8, 4.3]
   Posttest times [5.0, 5.0, 5.1, 4.9, 5.2, 4.5, 5.1, 4.9, 4.7, 4.4]

ANSWER FOR QUESTION 8

Seven Steps for Hypothesis Testing

STEP 1: Dependent t-test
STEP 2: Hypotheses $H_o: \mu_D \geq 0$ $H_1: \mu_D < 0$
STEP 3: Alpha Level (α)   Alpha level is 0.01.
STEP 4: Critical Value (C.V.)   C.V. = −2.821
STEP 5: Statistical Test   t = 1.899
STEP 6: Do Not Reject Null Hypothesis
STEP 7: Interpretation   *Summary*   There is not a statistically significant mean difference between the pretest and posttest 40-meter dash levels of the athletes who used supplements. The posttest levels are not significantly lower (t, (df = 9) = 1.899, p > 0.05).

9. A researcher would like to test the effectiveness of a drug that that affects weight levels. At the significance level of 0.05, determine whether the drug has an effect on weight. Here are the weights of the subjects before and after:

Pretest weight [190, 180, 200, 260, 210, 185, 187, 213, 145, 167]
Posttest weight [189, 181, 198, 259, 200, 184, 192, 210, 155, 166]

ANSWER FOR QUESTION 9

Seven Steps for Hypothesis Testing

STEP 1: Dependent t-test
STEP 2: Hypotheses $H_o: \mu_D = 0$ $H_1: \mu_D \neq 0$
STEP 3: Alpha Level (α)   Alpha level is 0.05.
STEP 4: Critical Value (C.V.)   C.V. = +/−1.833
STEP 5: Statistical Test   t = 0.829
STEP 6: Do Not Reject Null Hypothesis
STEP 7: Interpretation   *Summary*   There is not a statistically significant mean difference between the pretest and posttest weight levels after using the drug. The posttest levels are not significantly lower (t, (df = 9) = 0.829, p > 0.05).

10. A researcher would like to test whether there is a difference between a traditional sales program and an established sales program among a single group of subjects. At the significance level of 0.05, determine whether there is a difference. Here are the sales levels for the subjects when they used the traditional program and when they used the new program:

Traditional program [7, 9, 10, 8, 6]
New program [12, 10, 14, 17, 18]

ANSWER FOR QUESTION 10

Seven Steps for Hypothesis Testing

STEP 1: Dependent t-test
STEP 2: Hypotheses $H_o: \mu_D = 0$ $H_1: \mu_D \neq 0$
STEP 3: Alpha Level (α)   Alpha level is 0.05.
STEP 4: Critical Value (C.V.)   C.V. = +/–2.776
STEP 5: Statistical Test   t = 3.206
STEP 6: Reject Null Hypothesis
STEP 7: Interpretation   *Summary*   There is a statistically significant mean difference between the pretest and posttest sales levels after using the new program. The posttest levels are significantly higher (t, (df = 4) = 3.206, p > 0.05).

REFERENCES

Fisher, R. A. 1990. *Statistical methods, experimental design, and scientific inference*. Oxford: Oxford University Press.

Salsburg, D. 2001. *The lady tasting tea: How statistics revolutionized science in the twentieth century*. New York: W.H. Freeman.

The t Distribution

Table entries are values of t corresponding to proportions in one-tail or in two-tails combined.

One-Tail	0.25	0.10	0.05	0.025	0.01	0.005
Two-Tail	0.50	0.20	0.10	0.05	0.02	0.01
df = 1	1.000	3.078	6.314	12.71	31.82	63.66
2	0.816	1.886	2.920	4.303	6.965	9.925
3	0.765	1.638	2.353	3.182	4.541	5.841
4	0.741	1.533	2.132	2.776	3.747	4.604
5	0.727	1.476	2.015	2.571	3.365	4.032
6	0.718	1.440	1.943	2.447	3.143	3.707
7	0.711	1.415	1.895	2.365	2.998	3.499
8	0.706	1.397	1.860	2.306	2.896	3.355
9	0.703	1.383	1.833	2.262	2.821	3.250
10	0.700	1.372	1.812	2.228	2.764	3.169
11	0.697	1.363	1.796	2.201	2.718	3.106
12	0.695	1.356	1.782	2.179	2.681	3.055
13	0.694	1.350	1.771	2.160	2.650	3.012
14	0.692	1.345	1.761	2.145	2.624	2.977
15	0.691	1.341	1.753	2.131	2.602	2.947
16	0.690	1.337	1.746	2.120	2.583	2.921
17	0.689	1.333	1.740	2.110	2.567	2.898
18	0.688	1.330	1.734	2.101	2.552	2.878
19	0.688	1.328	1.729	2.093	2.539	2.861
20	0.687	1.325	1.725	2.086	2.528	2.845
21	0.686	1.323	1.721	2.080	2.518	2.831
22	0.686	1.321	1.717	2.074	2.508	2.819
23	0.685	1.319	1.714	2.069	2.500	2.807
24	0.685	1.318	1.711	2.064	2.492	2.797

Continued on next page.

National Science Teachers Association

Continued from previous page.

25	0.684	1.316	1.708	2.060	2.485	2.787
26	0.684	1.315	1.706	2.056	2.479	2.779
27	0.684	1.314	1.703	2.052	2.473	2.771
28	0.683	1.313	1.701	2.048	2.467	2.763
29	0.683	1.311	1.699	2.045	2.462	2.756
30	0.683	1.310	1.697	2.042	2.457	2.750
40	0.681	1.303	1.684	2.021	2.423	2.704
50	0.679	1.299	1.676	2.009	2.403	2.678
60	0.679	1.296	1.671	2.000	2.390	2.660
80	0.678	1.292	1.664	1.990	2.374	2.639
100	0.677	1.290	1.660	1.984	2.364	2.626
120	0.677	1.289	1.658	1.980	2.358	2.617
∞	0.674	1.282	1.645	1.960	2.326	2.576

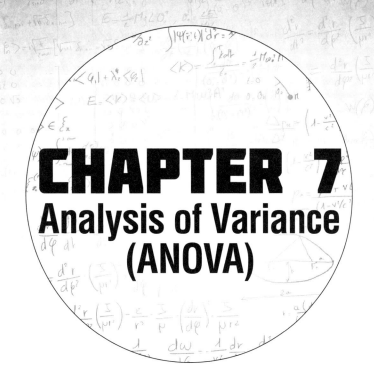

CHAPTER 7
Analysis of Variance (ANOVA)

OBJECTIVES

When you complete this chapter, you should be able to
1. demonstrate an understanding of the ANOVA to make inferences in classroom settings;
2. interpret ANOVA and its relevance to classroom settings;
3. compare ANOVA to other parametric procedures;
4. calculate ANOVAs from classroom and standardized test results; and
5. find, use, and apply ANOVA results.

Key Terms

When you complete this chapter, you should understand the following terms:
  degrees of freedom (df)
  F-ratios
  post-hoc tests

INTRODUCTION

Analysis of Variance (ANOVA) was first introduced by Sir Ronald Fisher in 1918 in his article, "Studies in Crop Variation." Subsequently, Fisher found new methods of analysis for researchers and gave researchers a tool to analyze data with more than two conditions or independent variables. In this chapter, we will explore ANOVA in its simplest form, one-way ANOVA, which is a statistical procedure that compares more than two means.

ONE-WAY ANOVA

The one-way ANOVA is used when a researcher is interested in testing the hypothesis of difference between more than two means. The ANOVA was developed to protect researchers from excessive risks of committing a type–I error, which was the case when researchers used multiple t-tests to analyze data from more than two means. ANOVA permits research-

used multiple t-tests to analyze data from more than two means. ANOVA permits researchers to use a single hypothesis test, decreasing the risk of committing a type-I error and increasing the level of confidence.

We will explore the one-way ANOVA and determine its calculation by following the Seven Steps for Hypothesis Testing. See Example 1.

> **Example 1:** A researcher is trying to determine whether there is a difference between the achievement scores of three small groups of students at a high school.
> Please use the $\alpha = .05$ alpha level.

Group 1	Group 2	Group 3
0	6	6
4	8	5
0	5	9
1	4	4
0	2	6

To begin the calculation, we must follow a sequence of steps that will be used throughout hypothesis testing. As you will see, the most general way to compute a one-way ANOVA is to define the statistical test statistic that can be calculated from scale of measurement and the collection of data. Below are the Seven Steps for Hypothesis Testing, which will be used to perform the calculation.

Seven Steps for Hypothesis Testing

STEP 1: Identify the appropriate statistical test to be used with the given data for analysis.

STEP 2: Hypotheses
- H_o (Null Hypothesis): This hypothesis proposes that no statistical significance exists. The null hypothesis attempts to show that no variation exists between the means, or that a single variable is not different. When the null hypothesis is not rejected, it is presumed to be true until statistical evidence nullifies it for an alternative hypothesis.
- H_1 (Alternative Hypothesis): The alternative hypothesis is the hypothesis that is contrary to the null hypothesis. This hypothesis should correspond with the researcher's assumption and is used to discern the conclusion.

STEP 3: Alpha Level (α)
Alpha level is the probability of making a type-I error (rejecting the null hypothesis when the null hypothesis is true). The lower this value is, the less chance that the researcher has of committing a type-I error. Also, an experiment should use a sample size large enough (i.e., $n = 20$) so that there is adequate statistical power. As a general rule, statistical analysis in education uses alpha levels of 0.05 or sometimes 0.01.

STEP 4: Critical Value (C.V.)
A number that causes rejection of the null hypothesis if a given test statistic is this number or more, and nonrejection of the null hypothesis if the test statistic is smaller than this number. The critical value can be found by analyzing a statistical table related to the appropriate statistical test.

STEP 5: Statistical Test
A statistical test provides a mechanism for making quantitative decisions. The intent is to determine whether there is enough evidence to "reject" the null hypothesis. Not rejecting may be a good result if we want to continue to act as if we "believe" the null hypothesis is true. Or it may be a disappointing result, possibly indicating we may not yet have enough data to "prove" something by rejecting the null hypothesis.

STEP 6: Reject or Do Not Reject the Null Hypothesis
By comparing the critical value to the calculated statistics, if the statistic exceeds the critical value, the null hypothesis is rejected.

STEP 7: Interpretation
Rejecting the null hypothesis may disclose that an independent variable/treatment did affect the dependent variable. However, not rejecting the null hypothesis may be a good result if we are looking for a result where no difference is desirable. Or it may be a disappointing result, possibly indicating we may not yet have enough data to "prove" something by rejecting the null hypothesis.

After understanding that the data must be analyzed by using a one-way ANOVA, we can proceed with the hypothesis.

STEP 1: Identify the Statistical Test
One-Way ANOVA

Most researchers test the null hypothesis (H_o), which is a statement that there is "no difference" between the sample mean and the population mean. Conversely, the alternative hypothesis (H_1) indicates that the sample mean and the population mean are not equal. In this case the hypotheses are written.

STEP 2: Hypotheses
H_o: $\mu_1 = \mu_2 = \mu_1 = \mu_3 ...$
H_1: At least one population mean is different from the others.

The third step is to set the alpha level (α). As explained earlier, the alpha level is used to determine the probability of committing a type-I error. Therefore, different studies have advantages or disadvantages when using higher or lower α-levels. Hence, you have to be careful about interpreting the meaning of these terms. For example, *higher* α-levels can *increase* the chance of a type-I error. Therefore, a lower α-level indicates that you are conducting a test that is more rigorous. Keep in mind that an alpha level of .05 means you have a 95% chance of saying there is no difference, when in fact the researcher is taking a risk of being wrong 5 in 100 times by rejecting the null hypothesis (committing a type-I error). However, an alpha level (α) of 0.01 means the researcher is being relatively cautious by minimizing the risk being wrong only 1 in 100 times to reject the null hypothesis. Based on the risk, the researcher decides the alpha level. In this case, the alpha level is set at 0.05.

> **STEP 3: Alpha Level**
> α = 0.05

The fourth step is to determine the critical value. The F-Table (p. 163) is used for the one-way ANOVA to establish the critical value. With ANOVA, since there is no direction, there is no need to determine a one-tail or a two-tailed test. In this case, the critical value is determined by intersecting the degrees of freedom with the probability value from the tails column. Since there are three groups with five subjects in each group, the degrees of freedom are determined by calculating the degrees of freedom between (numerator) and the degrees of freedom within (denominator). Hence, by using the F-Table, you can find the critical value by intersection of the degrees of freedom within and the degrees of freedom between. Another recommendation is to use a critical value calculator such as the one found here: *www.danielsoper.com/statcalc/calc04.aspx*.

> **STEP 4: Critical Value**
> d.f. between (numerator) = k (number of groups) = 3 – 1 = 2
> d.f. within (denominator) = (n–1) + (n–1) + (n–1) = (5–1) + (5–1) + (5–1) = 12
> d.f., 2, 12
> C.V. = +/–3.89

The fifth step is to compute the one-way ANOVA statistic. The formula for the ANOVA is:

$$F\text{-ratio} = \frac{MS\ (between)}{MS\ (within)}$$

In this equation, MS (between) estimates the variability of sample means in the population and the MS (within) estimates the variability of individual scores in the population. The fraction of the two scores is known as the F-ratio, which represents the Standard Error and estimates the amount of sampling error between the sample and the population. The MS is calculated by computing the ratio of the sum of the squares (SS) divided by the degrees of freedom (df).

STEP 5: Statistical Test (t-test)

Step #1

$$MS \text{ (between)} = \frac{SS \text{ (between)}}{df \text{ (between)}}$$

$$SS \text{ (between)} = \sum \frac{T^2}{n} - \frac{G^2}{N}$$

$$SS \text{ (between)} = \left(\frac{5^2}{5} + \frac{25^2}{5} + \frac{30^2}{5}\right) - \frac{60^2}{15} = 70$$

$(5 + 125 + 180) - 240 = 70$

T = each treatment or group total
G = the grand total, which combines all of the treatment Ts
N = the total number of scores
n = sample size of each group
df (between) = K − 1 = 3 − 1 = 2
K = Number of treatments.

$$MS(\text{between}) = \frac{SS \text{ (between)}}{df \text{ (between)}}$$

$$MS \text{ (between)} = \frac{70}{2} = 35$$

Step #2

$$MS \text{ (within)} = \frac{SS \text{ (within)}}{df \text{ (within)}}$$

SS (within) = Σ SS
SS (within) = 12+20+14 = 46

$SS_W = SS_T - SS_B$
$46 = 116 - 70$

df (within) = (n−1) + (n−1) +(n−1) = (5−1) + (5−1) +(5−1) = 12

Note: SS is calculated for each group by using the procedure that we used for calculating SS for standard deviation.

$$MS(\text{within}) = \frac{46}{12} = 3.83$$

Step #3

$$F\text{-ratio} = \frac{MS \text{ (between)}}{MS \text{ (within)}}$$

$$SS \text{ (between)} = \frac{35}{3.83} = 9.14$$

The calculation for the one-way ANOVA = 9.14, which measures the difference between the three groups.

The sixth step is to reject or to not reject the null hypothesis. This step takes into consideration a comparison of the critical value to the calculated F-ratio statistic, which compares the F-ratio of 9.14 to the C.V. of 3.88. Subsequently, since the F-ratio statistic exceeds the critical value of 3.88, the null hypothesis is rejected.

> **STEP 6: Reject or Do Not Reject the Null Hypothesis**
> Since the F-ratio of 9.14 is larger than C.V. of 3.88, we reject the null hypothesis.

With a one-way ANOVA, since there is a comparison of more than two groups, when the null hypothesis is rejected, we must use another statistical test known as a post-hoc test to determine where the difference or differences exist between the three groups. Without doing a post-hoc test, the null hypothesis is rejected without knowing how to interpret the alternative hypothesis. However, keep in mind that if the null hypothesis is not rejected, a post-hoc test is unnecessary.

POST-HOC TEST

Post-hoc tests are designed for when a researcher has already obtained a significant F-ratio with a one-way ANOVA. Subsequently, when three or more means are tested, the differences among means must be identified. The statistical test that we will use to identify differences is known as Scheffé's method, named after American statistician Henry Scheffé.

> Scheffé's method uses information that has been used in the original ANOVA calculation. Also, we use the same critical value. In this case, the critical value remains 3.88.

For each case, we must recalculate the MS (between), using only the two T values that are being compared.

$$\text{MS (between)} = \frac{\text{SS (between)}}{df \text{ (between)}}$$

Hence we must begin with SS (between).

First Comparison (Group 1 versus Group 3):

Step #1

$$\text{SS (between)} = \sum \frac{T^2}{n} - \frac{G^2}{N}$$

$$\text{SS (between)} = \frac{5^2}{5} + \frac{30^2}{5} - \frac{35^2}{10} = 62.5$$

Notice that we adjust the "G" to include the two groups that are being compared.

Step #2

$$\text{MS (between)} = \frac{\text{SS(between)}}{df \text{ (between)}}$$

$$\text{SS (between)} = \frac{62.50}{2} = 62.5$$

Notice that the df (between) is identical to the original df (between from the ANOVA calculation).

Step #3

$$\text{F-ratio} = \frac{\text{MS (between)}}{\text{MS (within)}}$$

$$\text{SS (between)} = \frac{31.25}{3.83} = 8.16$$

The MS (within) is the same as the MS (within) from the original ANOVA calculation. Since the F-ratio is higher than the C.V. of 3.88, we find that there is a significant difference between Group 1 and Group 3.

Second Comparison (Group 1 versus Group 2):

Step #1

$$\text{SS (between)} = \sum \frac{T^2}{n} - \frac{G^2}{N}$$

$$\text{SS (between)} = \frac{5^2}{5} + \frac{25^2}{5} - \frac{30^2}{10} = 40$$

Notice that we adjust the "G" to include the two groups that are being compared.

Step #2

$$\text{MS (between)} = \frac{\text{SS (between)}}{df\,(between)}$$

$$\text{MS (between)} = \frac{40}{2} = 20$$

Notice that the df (between) is identical to the original df (between from the ANOVA calculation).

Step #3

$$F\text{-ratio} = \frac{\text{MS (between)}}{\text{MS (within)}}$$

$$F\text{-ratio} = \frac{20}{3.83} = 5.22$$

The MS (within) is the same as the MS (within) from the original ANOVA calculation. Since the F-ratio is higher than the C.V. of 3.88, we find that there is a significant difference between Group #1 and Group #2.

Third Comparison (Group 2 versus Group 3):

Step #1

$$\text{SS (between)} = \sum \frac{T^2}{n} - \frac{G^2}{N}$$

$$\text{SS (between)} = \frac{25^2}{5} + \frac{30^2}{5} - \frac{55^2}{10} = 2.5$$

Notice that we adjust the "G" to include the two groups that are being compared.

Step #2

$$\text{MS (between)} = \frac{\text{SS (between)}}{df\,(between)}$$

$$\text{MS (between)} = \frac{2.5}{2} = 1.25$$

Notice that the df (between) is identical to the original df (between from the ANOVA calculation.

Step #3

$$F\text{-ratio} = \frac{\text{MS (between)}}{\text{MS (within)}}$$

$$F\text{-ratio} = \frac{1.25}{3.83} = .326$$

The MS (within) is the same as the MS (within) from the original ANOVA calculation. Since the F-ratio is not higher than the C.V. of 3.88, we find that there is not a significant difference between Group 1 and Group 2.

The seventh step is to interpret our findings. Since we rejected our null hypothesis, we conclude that the observed sample mean is significantly different from the population. Therefore, we conclude that there is a significant difference. After conducting Scheffé's method, we find that Group 3 is significantly higher than Group 1, Group 2 is significantly higher than Group 1, and that there is no difference between Group 2 and Group 3.

> **STEP 7: Interpretation**
> Rejecting the null hypothesis may disclose that an independent variable/treatment did affect the dependent variable.
>
> There is a significant difference showing that Group 3 is significantly higher than Group 1, Group 2 is significantly higher than Group 1, and that there is no difference between Group 2 and Group 3.

CALCULATOR EXPLORATION

Using the TI-83 or TI-84, you can easily compute a one-way ANOVA. Using the ANOVA test from Example 1, we will demonstrate how to calculate a one-way ANOVA with a graphing calculator.

Press [**EDIT**].
Press [**ENTER**].

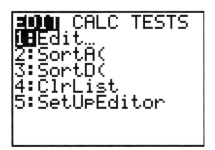

Enter relevant data in [L1], [L2], [L3].

Select [**TESTS H:ANOVA**].

```
EDIT CALC TESTS
B↑2-PropZInt…
C:χ²-Test…
D:χ²GOF-Test…
E:2-SampFTest…
F:LinRegTTest…
G:LinRegTInt…
H:ANOVA(
```

Enter [(L1, L2, L3)].

```
ANOVA(L1,L2,L3)
```

Press [**ENTER**].

```
One-way ANOVA
 F=9.130434783
 p=.0038886518
 Factor
  df=2
  SS=70
↓ MS=35
```

The one-way ANOVA statistic is 9.13. Since 9.13 > 3.88 and p-value is less than 0.05, we reject the null hypothesis.

SUMMARY

This chapter introduces the one-way ANOVA, which is a statistical tool for hypothesis testing. A one-way ANOVA is used to compare more than two samples to generalize to a population for which the means are known.

CHAPTER 7

The one-way ANOVA uses data from separate groups and focuses on a research question that attempts to evaluate the possibility of a significant mean difference. Hence, the one-way ANOVA is a tool for researchers to compare means and generalize findings to a population. In essence, a one-way ANOVA is a parametric procedure for determining whether significant differences exist.

Seven Steps for Hypothesis Testing

STEP 1: Identify the appropriate statistical test to be used with the given data for analysis.

STEP 2: Hypotheses
- H_0 (Null Hypothesis): This hypothesis proposes that no statistical significance exists. The null hypothesis attempts to show that no variation exists between the means, or that a single variable is not different. When the null hypothesis is not rejected, it is presumed to be true until statistical evidence nullifies it for an alternative hypothesis.
- H_1 (Alternative Hypothesis): The alternative hypothesis is the hypothesis that is contrary to the null hypothesis. This hypothesis should correspond with the researcher's assumption and is used to discern the conclusion.

STEP 3: Alpha Level (α)
Alpha level is the probability of making a type-I error (rejecting the null hypothesis when the null hypothesis is true). The lower this value is, the less chance that the researcher has of committing a type-I error. Also, an experiment should use a sample size large enough (i.e., $n = 20$) so that there is adequate statistical power. As a general rule, statistical analysis in education uses alpha levels of 0.05 or sometimes 0.01.

STEP 4: Critical Value (C.V.)
A number that causes rejection of the null hypothesis if a given test statistic is this number or more, and nonrejection of the null hypothesis if the test statistic is smaller than this number. The critical value can be found by analyzing a statistical table related to the appropriate statistical test.

STEP 5: Statistical Test
A statistical test provides a mechanism for making quantitative decisions. The intent is to determine whether there is enough evidence to "reject" the null hypothesis. Not rejecting may be a good result if we want to continue to act as if we "believe" the null hypothesis is true. Or it may be a disappointing result, possibly indicating we may not yet have enough data to "prove" something by rejecting the null hypothesis.

STEP 6: Reject or Do Not Reject the Null Hypothesis
By comparing the critical value to the calculated statistics, if the statistic exceeds the critical value, the null hypothesis is rejected.

CHAPTER 7

> **STEP 7: Interpretation**
> Rejecting the null hypothesis may disclose that an independent variable/treatment did affect the dependent variable. However, not rejecting the null hypothesis may be a good result if we are looking for a result where no difference is desirable. Or it may be a disappointing result, possibly indicating we may not yet have enough data to "prove" something by rejecting the null hypothesis.

CHAPTER REVIEW QUESTIONS

Use the Seven Steps for Hypothesis Testing to solve the following problems.

1. Determine whether there is a difference among three laboratory books used by biology students at the 0.05 level of significance. Below are the scores from the three samples:
   Group 1: 90, 100, 40, 80, 70
   Group 2: 40, 60, 80, 20, 100
   Group 3: 10, 30, 40, 50, 20

CHAPTER 7

ANSWER FOR QUESTION 1

Seven Steps for Hypothesis Testing

STEP 1: One-Way ANOVA
STEP 2: Hypotheses H_o: $\mu_1 = \mu_2 = \mu_1 = \mu_3 \ldots$ H_1: At least one population mean is different from the others.
STEP 3: Alpha Level (α)   Alpha level is 0.05.
STEP 4: Critical Value (C.V.)   C.V. = 3.88
STEP 5: Statistical Test   F-ratio = 4.596
STEP 6: Reject Null Hypothesis
STEP 7: Interpretation   *Summary*   There is a statistically significant mean difference between the sample and the population (F= 4.596, p < 0.05). Post Hoc tests show that there is a significant difference between Group 1 and Group 3, with Group 1 being significantly greater.

2. At the 0.05 level of significance, determine if there is a difference among the anxiety scores of students enrolled in three levels of science courses. Below are the scores from the three samples:
   Group 1 (level 1): 90, 75, 60, 50, 95, 64
   Group 2 (level 2): 81, 66, 72, 80, 74, 84
   Group 3 (level 3): 82, 84, 83, 94, 87, 79

ANSWER FOR QUESTION 2

Seven Steps for Hypothesis Testing

STEP 1: One-Way ANOVA
STEP 2: Hypotheses H_o: $\mu_1 = \mu_2 = \mu_1 = \mu_3 \ldots$ H_1: at least one population mean is different from the others.
STEP 3: Alpha Level (α)   Alpha level is 0.05.
STEP 4: Critical Value (C.V.)   C.V. = 3.68
STEP 5: Statistical Test   F-ratio = 1.929
STEP 6: Do Not Reject Null Hypothesis
STEP 7: Interpretation   *Summary*   There is no statistically significant mean difference between the sample and the population (F = 1.929, p > 0.05).

3. A scientist examines the effect of different music levels on anxiety. At the 0.05 level, determine whether there is a difference. Below are the scores from the three samples:
   Group 1 (Loud): 9, 5, 11, 2, 6, 7
   Group 2 (Medium): 8, 13, 13, 9, 10, 18
   Group 3 (Soft): 16, 15, 13, 7, 9, 5

CHAPTER 7

ANSWER FOR QUESTION 3

Seven Steps for Hypothesis Testing

STEP 1: One-Way ANOVA
STEP 2: Hypotheses $H_o: \mu_1 = \mu_2 = \mu_1 = \mu_3 ...$ $H_1:$ at least one population mean is different from the others.
STEP 3: Alpha Level (α)   Alpha level is 0.05
STEP 4: Critical Value (C.V.)   C.V. = 3.68
STEP 5: Statistical Test   F-ratio = 3.114
STEP 6: Do Not Reject Null Hypothesis
STEP 7: Interpretation   *Summary*   There is no statistically significant mean difference between the sample and the population (F = 3.114, p > 0.05).

4. A researcher wants to determine if the sales levels are different in three geographic regions. At the 0.05 level, determine whether there is a difference. Below are the scores from the three samples:
   Group 1 (West): 4, 6, 2, 9, 11, 4
   Group 2 (Central): 13, 12, 8, 17, 9, 10
   Group 3 (East): 6, 8, 4, 15, 18, 17

ANSWER FOR QUESTION 4

Seven Steps for Hypothesis Testing

STEP 1: One-Way ANOVA
STEP 2: Hypotheses Ho: $\mu_1 = \mu_2 = \mu_1 = \mu_3 ...$ H_1: At least one population mean is different from the others.
STEP 3: Alpha Level (α) Alpha level is 0.05.
STEP 4: Critical Value (C.V.) C.V. = 3.68
STEP 5: Statistical Test F-ratio = 2.988
STEP 6: Do Not Reject Null Hypothesis
STEP 7: Interpretation *Summary* There is no statistically significant mean difference between the sample and the population (F = 2.988, $p > 0.05$).

5. A study is designed to examine the cholesterol levels of subjects in three different age groups. At the 0.05 level, determine whether there is a difference. Below are the scores from the three samples:
   Group 1 (10 to 20 years): 150, 160, 155, 174, 140
   Group 2 (21 to 40 years): 192, 184, 193, 210, 240
   Group 3 (41 to 60 years): 184, 245, 221, 201, 295

ANSWER FOR QUESTION 5

Seven Steps for Hypothesis Testing

STEP 1: One-Way ANOVA
STEP 2: Hypotheses   Ho: $\mu_1 = \mu_2 = \mu_1 = \mu_3 ...$ H_1: at least one population mean is different from the others.
STEP 3: Alpha Level (α)   Alpha level is 0.05.
STEP 4: Critical Value (C.V.)   C.V. = 3.88
STEP 5: Statistical Test   F-ratio = 8.242
STEP 6: Do Not Reject Null Hypothesis
STEP 7: Interpretation   *Summary*   There is a statistically significant mean difference between the sample and the population (F = 8.242, p < 0.05). There is a significant difference between Group 1 and Group 3, with Group 3 being significantly higher than Group 1.

6. An automotive engineer is testing the 0-60 mph times of three versions of a car. At the 0.05 level, determine whether there is a difference. Below are the scores from the three samples:
   Version 1: 8.9, 6.8, 8.1, 7.4, 6.2
   Version 2: 14.6, 9.5, 8.2, 7.8, 8.8,
   Version 3: 5.9, 6.2, 6.7, 5.8, 6.3

ANSWER FOR QUESTION 6

Seven Steps for Hypothesis Testing

STEP 1: One-Way ANOVA
STEP 2: Hypotheses Ho: $\mu_1 = \mu_2 = \mu_1 = \mu_3 \ldots$ H_1: At least one population mean is different from the others.
STEP 3: Alpha Level (α) Alpha level is 0.05.
STEP 4: Critical Value (C.V.) C.V. = 3.88
STEP 5: Statistical Test F-ratio = 5.585
STEP 6: Reject Null Hypothesis
STEP 7: Interpretation *Summary* There is a statistically significant mean difference between the sample and the population (F = 5.585, p < 0.05). There is a significant difference between Group 2 and Group 3, with Group 2 being significantly higher than Group 3.

7. A psychologist is looking at the productivity levels of employees who are grouped according to their measure states of three different anxiety levels during work hours. At the 0.05 level, determine whether there is a difference. Below are the scores from the three samples:
   Sample 1: 8, 6, 8, 7, 6
   Sample 2: 14, 9, 8, 7, 8
   Sample 3: 5, 6, 6, 5, 6

CHAPTER 7

ANSWER FOR QUESTION 7

Seven Steps for Hypothesis Testing

STEP 1: One-Way ANOVA
STEP 2: Hypotheses $H_o: \mu_1 = \mu_2 = \mu_1 = \mu_3 \ldots$ H_1: at least one population mean is different from the others.
STEP 3: Alpha Level (α)   Alpha level is 0.05.
STEP 4: Critical Value (C.V.)   C.V. = 3.88
STEP 5: Statistical Test   F-ratio = 5.489
STEP 6: Reject Null Hypothesis
STEP 7: Interpretation   *Summary*   There is a statistically significant mean difference between the sample and the population ($F = 5.489$, $p < .05$). There is a significant difference between Group 2 and Group 3, with Group 2 being significantly higher than Group 3.

8. A social scientist is looking at the methods to enhance worker job satisfaction levels of teachers with three different treatments. At the 0.05 level, determine whether there is a difference. Below are the scores from the three samples:
   Group 1 (Lower class sizes): 8, 6, 8, 7, 6
   Group 2 (Fewer work hours): 14, 9, 8, 7, 8
   Group 3 (Higher pay): 5, 6, 6, 5, 6

ANSWER FOR QUESTION 8

Seven Steps for Hypothesis Testing

STEP 1: One-Way ANOVA
STEP 2: Hypotheses H_o: $\mu_1 = \mu_2 = \mu_1 = \mu_3 \ldots$ H_1: At least one population mean is different from the others.
STEP 3: Alpha Level (α)   Alpha level is 0.05.
STEP 4: Critical Value (C.V.)   C.V. = 3.88
STEP 5: Statistical Test   F-ratio = 32.386
STEP 6: Reject Null Hypothesis
STEP 7: Interpretation   *Summary*   There is a statistically significant mean difference between the sample and the population (F = 32.386, p <.05). There is a significant difference between Group 2 and Group 3, with Group 3 being significantly higher than Group 2 and a significant difference between Group 1 and Group 2, with Group 2 being significantly higher than Group 1.

9. A research looked at the job stability of people who have different relationship levels. At the 0.05 level, determine whether there is a difference. Below are the scores from the three samples:
   Group 1 (Lower): 5, 1, 0, 8, 3
   Group 2 (Medium): 0, 6, 9, 2, 7
   Group 3 (High): 8, 9, 1, 3, 8

ANSWER FOR QUESTION 9

Seven Steps for Hypothesis Testing

STEP 1: One-Way ANOVA
STEP 2: Hypotheses H_0: $\mu_1 = \mu_2 = \mu_1 = \mu_3 \ldots$ H_1: At least one population mean is different from the others.
STEP 3: Alpha Level (α)   Alpha level is 0.05.
STEP 4: Critical Value (C.V.)   C.V. = 3.88
STEP 5: Statistical Test   F-ratio = 0.5940
STEP 6: Do Not Reject Null Hypothesis
STEP 7: Interpretation   **Summary**   There is not a statistically significant mean difference between the sample and the population (F = 0.5940, $p < 0.05$).

10. A researcher is trying to determine if color has an effect on mood among three different groups. At the 0.05 level, determine whether there is a difference. Below are the scores from the three samples:
    Group 1 (Black): 2, 3, 2, 4, 1
    Group 2 (Grey): 4, 5, 3, 4, 8
    Group 3 (Green): 7, 8, 6, 8, 10

ANSWER FOR QUESTION 10

Seven Steps for Hypothesis Testing

STEP 1: One-Way ANOVA
STEP 2: Hypotheses H_o: $\mu_1 = \mu_2 = \mu_1 = \mu_3 \ldots$ H_1: At least one population mean is different from the others.
STEP 3: Alpha Level (α)   Alpha level is 0.05.
STEP 4: Critical Value (C.V.)   C.V. = 3.88
STEP 5: Statistical Test   F-ratio = 15.251
STEP 6: Reject Null Hypothesis
STEP 7: Interpretation   *Summary*   There is a statistically significant mean difference between the sample and the population (F = 15.251, p < 0.05). There is a significant difference between Group 1 and Group 3, with Group 3 being significantly higher than Group 1 and a significant difference between Group 2 and Group 3, with Group 3 being significantly higher than Group 2.

REFERENCES

Fisher, R. A. 1990. *Statistical methods, experimental design, and scientific inference.* Oxford: Oxford University Press.

Salsburg, D. 2001. *The lady tasting tea: How statistics revolutionized science in the twentieth century.* New York: W.H. Freeman.

F-Table
alpha=.05

		NUMERATOR (Between)															
Df-Denominator (Within)	df	1	2	3	4	5	6	7	8	9	10	12	15	20	24	30	40	60	120	inf
1		161.4	199.5	215.7	224.6	230.2	234.0	236.8	238.9	240.5	241.9	243.9	245.9	248.0	249.1	250.1	251.1	252.2	253.3	254.3
2		18.51	19.00	19.16	19.25	19.30	19.33	19.35	19.37	19.38	19.40	19.41	19.43	19.45	19.45	19.46	19.47	19.48	19.49	19.50
3		10.13	9.55	9.28	9.12	9.01	8.94	8.89	8.85	8.81	8.79	8.74	8.70	8.66	8.64	8.62	8.59	8.57	8.55	8.53
4		7.71	6.94	6.59	6.39	6.26	6.16	6.09	6.04	6.00	5.96	5.91	5.86	5.80	5.77	5.75	5.72	5.69	5.66	5.63
5		6.61	5.79	5.41	5.19	5.05	4.95	4.88	4.82	4.77	4.74	4.68	4.62	4.56	4.53	4.50	4.46	4.43	4.40	4.36
6		5.99	5.14	4.76	4.53	4.39	4.28	4.21	4.15	4.10	4.06	4.00	3.94	3.87	3.84	3.81	3.77	3.74	3.70	3.67
7		5.59	4.74	4.35	4.12	3.97	3.87	3.79	3.73	3.68	3.64	3.57	3.51	3.44	3.41	3.38	3.34	3.30	3.27	3.23
8		5.32	4.46	4.07	3.84	3.69	3.58	3.50	3.44	3.39	3.35	3.28	3.22	3.15	3.12	3.08	3.04	3.01	2.97	2.93
9		5.12	4.26	3.86	3.63	3.48	3.37	3.29	3.23	3.18	3.14	3.07	3.01	2.94	2.90	2.86	2.83	2.79	2.75	2.71
10		4.96	4.10	3.71	3.48	3.33	3.22	3.14	3.07	3.02	2.98	2.91	2.85	2.77	2.74	2.70	2.66	2.62	2.58	2.54
11		4.84	3.98	3.59	3.36	3.20	3.09	3.01	2.95	2.90	2.85	2.79	2.72	2.65	2.61	2.57	2.53	2.49	2.45	2.40
12		4.75	3.89	3.49	3.26	3.11	3.00	2.91	2.85	2.80	2.75	2.69	2.62	2.54	2.51	2.47	2.43	2.38	2.34	2.30
13		4.67	3.81	3.41	3.18	3.03	2.92	2.83	2.77	2.71	2.67	2.60	2.53	2.46	2.42	2.38	2.34	2.30	2.25	2.21
14		4.60	3.74	3.34	3.11	2.96	2.85	2.76	2.70	2.65	2.60	2.53	2.46	2.39	2.35	2.31	2.27	2.22	2.18	2.13
15		4.54	3.68	3.29	3.06	2.90	2.79	2.71	2.64	2.59	2.54	2.48	2.40	2.33	2.29	2.25	2.20	2.16	2.11	2.07
16		4.49	3.63	3.24	3.01	2.85	2.74	2.66	2.59	2.54	2.49	2.42	2.35	2.28	2.24	2.19	2.15	2.11	2.06	2.01
17		4.45	3.59	3.20	2.96	2.81	2.70	2.61	2.55	2.49	2.45	2.38	2.31	2.23	2.19	2.15	2.10	2.06	2.01	1.96
18		4.41	3.55	3.16	2.93	2.77	2.66	2.58	2.51	2.46	2.41	2.34	2.27	2.19	2.15	2.11	2.06	2.02	1.97	1.92
19		4.38	3.52	3.13	2.90	2.74	2.63	2.54	2.48	2.42	2.38	2.31	2.23	2.16	2.11	2.07	2.03	1.98	1.93	1.88
20		4.35	3.49	3.10	2.87	2.71	2.60	2.51	2.45	2.39	2.35	2.28	2.20	2.12	2.08	2.04	1.99	1.95	1.90	1.84

Continued on next page.

Continued from previous page.

21	4.32	3.47	3.07	2.84	2.68	2.57	2.49	2.42	2.37	2.32	2.25	2.18	2.10	2.05	2.01	1.96	1.92	1.87	1.81
22	4.30	3.44	3.05	2.82	2.66	2.55	2.46	2.40	2.34	2.30	2.23	2.15	2.07	2.03	1.98	1.94	1.89	1.84	1.78
23	4.28	3.42	3.03	2.80	2.64	2.53	2.44	2.37	2.32	2.27	2.20	2.13	2.05	2.01	1.96	1.91	1.86	1.81	1.76
24	4.26	3.40	3.01	2.78	2.62	2.51	2.42	2.36	2.30	2.25	2.18	2.11	2.03	1.98	1.94	1.89	1.84	1.79	1.73
25	4.24	3.39	2.99	2.76	2.60	2.49	2.40	2.34	2.28	2.24	2.16	2.09	2.01	1.96	1.92	1.87	1.82	1.77	1.71
26	4.23	3.37	2.98	2.74	2.59	2.47	2.39	2.32	2.27	2.22	2.15	2.07	1.99	1.95	1.90	1.85	1.80	1.75	1.69
27	4.21	3.35	2.96	2.73	2.57	2.46	2.37	2.31	2.25	2.20	2.13	2.06	1.97	1.93	1.88	1.84	1.79	1.73	1.67
28	4.20	3.34	2.95	2.71	2.56	2.45	2.36	2.29	2.24	2.19	2.12	2.04	1.96	1.91	1.87	1.82	1.77	1.71	1.65
29	4.18	3.33	2.93	2.70	2.55	2.43	2.35	2.28	2.22	2.18	2.10	2.03	1.94	1.90	1.85	1.81	1.75	1.70	1.64
30	4.17	3.32	2.92	2.69	2.53	2.42	2.33	2.27	2.21	2.16	2.09	2.01	1.93	1.89	1.84	1.79	1.74	1.68	1.62
40	4.08	3.23	2.84	2.61	2.45	2.34	2.25	2.18	2.12	2.08	2.00	1.92	1.84	1.79	1.74	1.69	1.64	1.58	1.51
60	4.00	3.15	2.76	2.53	2.37	2.25	2.17	2.10	2.04	1.99	1.92	1.84	1.75	1.70	1.65	1.59	1.53	1.47	1.39
120	3.92	3.07	2.68	2.45	2.29	2.17	2.09	2.02	1.96	1.91	1.83	1.75	1.66	1.61	1.55	1.50	1.43	1.35	1.25
∞	3.84	3.00	2.60	2.37	2.21	2.10	2.01	1.94	1.88	1.83	1.75	1.67	1.57	1.52	1.46	1.39	1.32	1.22	1.00

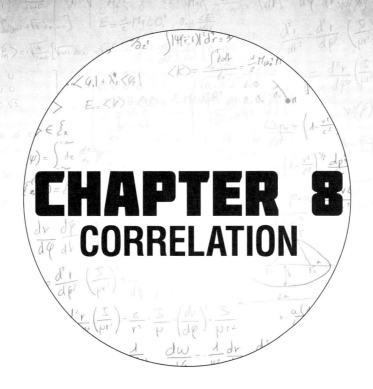

CHAPTER 8
CORRELATION

OBJECTIVES

When you complete this chapter, you should be able to
1. recognize the strength and type of relationship between two variables,
2. demonstrate an understanding of a correlation coefficient,
3. select and use appropriate statistical methods to compute a correlation coefficient,
4. read and interpret a scatter plot, and
5. compute Pearson and Spearman rank-order correlation coefficients.

Key Terms

When you complete this chapter, you should understand the following terms
correlation coefficient
    Pearson correlation coefficient
    relationship
    scatter plot
    Spearman rank-order correlation coefficient

Thus far, we have explored individual statistics, which illustrate individual sets of scores through frequencies, central tendency, and variance. Sometimes, however, you will find it necessary to determine the relationship between two sets of scores. For example, are high SAT verbal scores (now called critical reading scores) associated with high college English grades? Do high-ability students tend to excel in all of their academic subjects? This chapter will discuss the concept of *correlation*, which is used in later chapters that will explain the concepts of validity and reliability. Here, we introduce the **Pearson correlation coefficient**, a statistic that is used with ratio or interval-scaled data. In addition, we introduce the Spearman correlation, which is used with ranked or ordinal scaled data.

THE PEARSON CORRELATION COEFFICIENT

The **correlation coefficient** is a statistic that illustrates the existence of a linear **relationship** between two variables. It also expresses the strength of the relationship. For example, you might ask the question, are higher teacher salaries linked to higher academic achievement scores among students in a school district?

The numerical values of a correlation coefficient range between –1.00 and +1.00. The higher the numerical value, the stronger the relationship between the two variables (see Table 8.1). If the coefficient has a positive sign, the relationship is positive: If one value is high, the other value is high. Conversely, if the coefficient is negative, the relationship is negative: If one value is high, the other value is low. For example, when a correlation coefficient is positive, if x is high, so is y. You would expect that a person who scores high on an algebra aptitude test would also obtain high grades in algebra. The correlation is also positive when a low value for x corresponds to a low value for y.

Conversely, if a correlation is negative, high scores on x are associated with low scores on y.

As an example, let's look at the relationship between the SAT math scores and the trigonometry grades of 10th, 11th, and 12th graders. Table 8.2 shows these students' SAT mathematics scores and trigonometry grades. To determine the correlation coefficient, we will calculate the Pearson correlation coefficient for this set of scores.

The Pearson correlation coefficient, sometimes called the Pearson product-moment correlation, is a measure of the linear relationship between the paired values of two variables (x and y). The equation takes into account each paired value and uses the mean, standard deviation, and z-score formulas in its computation (see Equation 8.1).

Table 8.1

Interpretations of Correlations

Correlation	Interpretation
0.000 to 0.200	Very Weak
0.201 to 0.400	Weak
0.401 to 0.600	Moderate
0.601 to 0.800	Strong
0.801 to 1.000	Very Strong

Note: Numerical values can be plus or minus.

Table 8.2

SAT Math Scores and Trigonometry Achievement Scores

Forrest	740	95
Jamal	680	90
Alexandra	660	90
Lauren	550	86
Sabrina	500	80
Roberto	480	80
Chang	500	75
Jane	470	75
Rick	480	70
Pete	400	65

Equation 8.1

$$r = \frac{\sum (x - \bar{x})(y - \bar{y})}{(n-1) S_x S_y}$$

In this equation, $(x - \bar{x})$ is the deviation of the x variable from the mean, $(y - \bar{y})$ is the deviation of the y variable from the mean, S_x is the sample standard deviation for the x variable, S_y is the sample standard deviation for the y variable, and n represents the number of pairs of scores. To simplify this equation, we will represent r as the average value of the products of paired z-scores. This formula is found in Equation 8.2.

Equation 8.2

$$r = \frac{\sum z_x z_y}{n - 1}$$

As our first example, we will examine the relationship between SAT mathematics scores and the percentage scores from a 12th-grade trigonometry class. Table 8.3 shows the relationship between the two variables, using the definitional formula given in Equation 8.2.

Table 8.3

Pearson Correlation Calculation

Student	SAT (x)	Trig. Grade (y)	$z_x = (x - \bar{x})/S_x$	$z_y = (y - \bar{y})/S_y$	$z_x z_y$
Forrest	740	95	1.77	1.50	2.66
Jamal	680	90	1.22	0.98	1.20
Alexandra	660	90	1.04	0.98	1.02
Lauren	550	86	0.04	-0.57	0.02
Sabrina	500	80	-0.42	-0.05	0.02
Rob	480	80	-0.60	-0.05	0.03
Chang	500	75	-0.42	-0.57	0.24
Jane	470	75	-0.69	-0.67	0.46
Rick	480	70	-0.60	-1.08	0.65
Pete	400	65	-1.33	-1.60	2.13
	$\bar{x} = 546$ $S_x = 109.87$	$\bar{y} = 80.50$ $S_y = 9.69$			$\sum z_x z_y = 8.43$

Notice that, for each variable (x and y), z-scores are calculated to transform the data onto the same scale. Next z_x is multiplied by z_y and then summed with the formula $Sz_x z_y$, which will total to a positive sum value when a majority of positive z_x scores is multiplied by positive z_y scores, or, in a negative sum value, when a majority of positive z_x scores is multiplied by negative z_y scores. Calculation 8.1 shows that $r = 0.94$, which translates to a very strong positive correlation between SAT math scores and trigonometry scores. There-

CHAPTER 8

fore, we can conclude that high SAT scores are positively related to high trigonometry scores.

$$r = \frac{\sum z_x z_y}{n-1}$$

or

$$r = \frac{8.43}{10-1} = 0.94$$

Calculation 8.1

CALCULATOR EXPLORATION

Using either the TI-73, TI-83, or TI-84 graphing calculator, you can easily compute a correlation coefficient by following these keystrokes.

TI-73

Step 1. Display list editor by pressing [LIST]. Here, you can enter up to 999 elements. Under [L1], list all of the scores from the SATs that are listed in Table 8.3. Next, under [L2], list all of the scores listed under trigonometry grades in Table 8.3.

Step 2. Press [2nd LIST] to activate [STAT]. Next, scroll to the right and select [CALC]. Next, scroll down to [5: LinReg (ax+b)], press [5], then press [ENTER].

Step 3. Press [2nd APPS] to activate [VARS]. Next, scroll down to [3: Statistics], and press [3]. Next, scroll over to [EQ] and scroll down to [5: r], select [5], and press [ENTER]. Press [ENTER] again to get the correlation coefficient.

TI-83/TI-84

Step 1. First press [STAT]. Next, select [1: Edit] by pressing [1]. Here you can enter up to 999 elements. Under [L1], list all of the scores from the SATs that are listed in Table 8.3. Next, under [L2], list all of the scores listed under trigonometry grades in Table 8.3.

Step 2. Press [STAT] and scroll to the right and select [CALC]. Next, scroll down to [4: LinReg (ax+b)], press [4], and press [ENTER].

Step 3. Press [VARS]. Next, scroll down to [5: Statistics], and press [5]. Next, scroll over to [EQ]. Next, scroll down to [7: r] and press [7]. [ENTER] again to get the validity coefficient.

CORRELATION AND CAUSE AND EFFECT

People often think that correlation means the same thing as causation. Although a correlation means there is a relationship between two variables, it does not mean that one causes the other. For example, there is a positive correlation between ice cream consumption and death by drowning. However, common sense tells us that eating ice cream does not cause death by drowning; simply, when the weather is hot, more people eat ice cream and more people go swimming! As a further example, a plot of monthly sales of ice cream against monthly deaths from heart disease would show a negative correlation. Again, based on hundreds of research studies, it is hardly likely that eating ice cream protects from heart disease. It is simply that the mortality rate from heart disease is inversely related—and ice cream consumption positively related—to a third factor: environmental temperature. It is important to understand that a high correlation between two variables does not imply that one causes the other.

SCATTER PLOTS

The relationship between two variables can be illustrated in a ***scatter plot***, which is a graph showing the paired scores. Figure 8.1 shows a scatter plot for SAT mathematics section scores and trigonometry grades. Notice that, because our relationship is very close to a perfect +1.0, it is almost a straight line.

CHAPTER 8

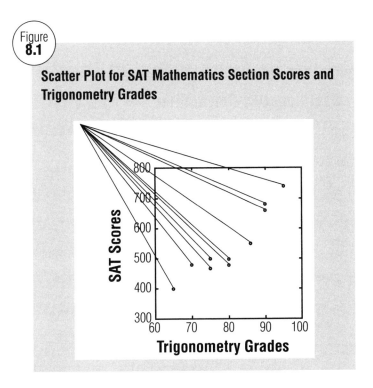

Figure 8.1

Scatter Plot for SAT Mathematics Section Scores and Trigonometry Grades

Figure 8.2 is a scatter plot of a perfect +1.00 correlation. In this case, we have positive scores for variables *x* and *y*. Likewise, since variable *x* is Fahrenheit temperature and variable *y* is the corresponding centigrade temperature, this scatter plot shows a perfectly straight line, which is known as a regression line.

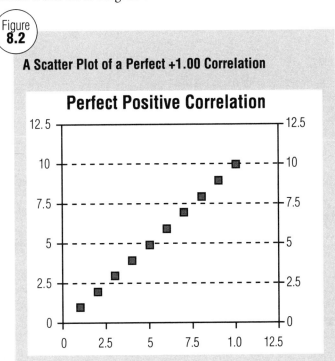

Figure 8.2

A Scatter Plot of a Perfect +1.00 Correlation

CHAPTER 8

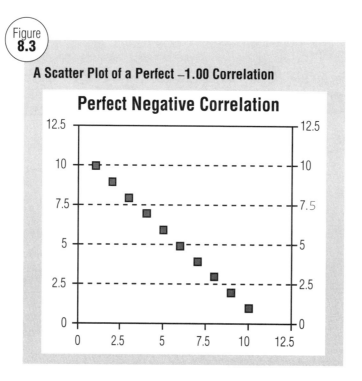

Figure 8.3

A Scatter Plot of a Perfect −1.00 Correlation

Figure 8.3 shows a scatter plot of a perfect −1.00 correlation. It shows the average velocity (x) to the race time (y) of a person running a 5k race, we would find that as the runner's running speed increases, his or her time decreases, giving us a perfect negative correlation.

Figure 8.4 shows a scatter plot of a high positive correlation between people's heights (x) and weights (y).

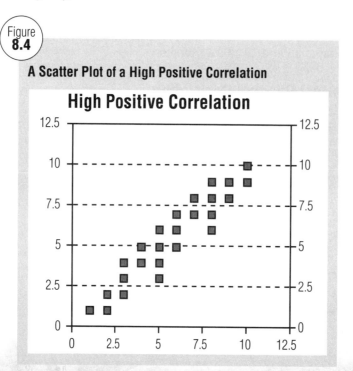

Figure 8.4

A Scatter Plot of a High Positive Correlation

Figure 8.5 shows a scatter plot of a low negative correlation. The scatter plot compares the amount of money spent on education (x) with test scores (y), showing that as spending increases slightly, academic achievement slightly decreases, giving us a low negative correlation between the two variables (Christmann and Badgett 2000). It is important to remember, however, that a correlation shows only that two variables are related; it does not show a cause and effect relationship. This happens because slight increases in teachers' pay do not affect how hard teachers work—which would often result in higher academic achievement. In essence, most teachers are dedicated professionals who put forth great effort regardless of pay increases.

Figure 8.6 is a scatter plot that shows no correlation. For example, if we were to compare combined SAT scores (x) to the heights (y) of high school seniors, we would find that there is no relationship between the two variables, which should yield a correlation of 0, for no correlation.

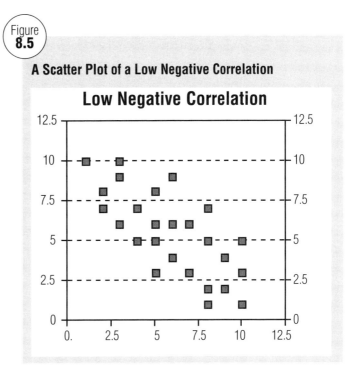

Figure 8.5

A Scatter Plot of a Low Negative Correlation

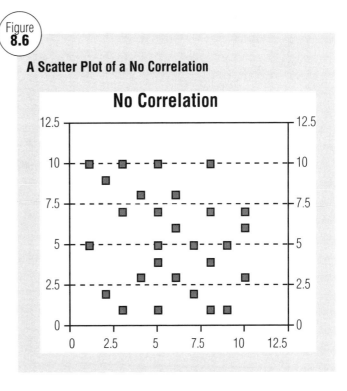

Figure 8.6

A Scatter Plot of a No Correlation

Check for Understanding

8.1. Which correlation coefficient shows the weakest relationship?
   a. + 0.94
   b. –0.77
   c. + 0.15
   d. –0.22

8.2. Calculate a Pearson correlation coefficient between the two variables given in the table to the right.

8.3. How would the calculated correlation coefficient from question 5.2 be classified according to the categories given in Table 5.1?

(x)	(y)
100	77
95	76
90	80
85	70
80	75
75	74
70	72
65	75
60	69
55	60

8.4. How would you best describe the scatter plot below?
   a. A scatter plot with no correlation
   b. A perfect positive correlation
   c. A low negative correlation
   d. A high positive correlation

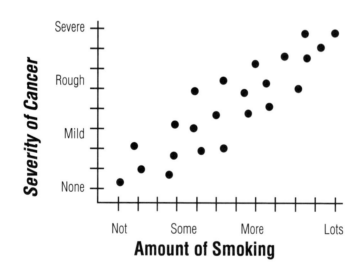

SPEARMAN RANK-ORDER CORRELATION COEFFICIENT

The ***Spearman rank-order correlation coefficient*** describes the linear relationship between two variables that are ranked on an ordinal scale of measurement. For example, you could use the Spearman rank-order coefficient to show such things as academic achievement based on class rank and IQ.

The symbol r_s represents the Spearman correlation coefficient (see Equation 8.3). Just like the Pearson r correlation coefficient, the Spearman correlation ranges between -1.00 and +1.00. We interpret the strength of a Spearman rank-order correlation the same way we interpret the strength of the correlations in Table 8.1.

Equation 8.3
$$r_s = 1 - \frac{6 \Sigma D^2}{n(n^2 - 1)}$$

For example, suppose we are interested in whether 12th-grade students' GPAs are related to their rankings on the SAT (see Table 8.4, p. 176). The first step in calculating a Spearman rank-order correlation, as shown in Table 8.4, is to rank the GPAs and the SAT scores. Once the scores have been ranked, we subtract the difference between the ranks, D. Next we square the differences (D^2) and compute the sum of the squared differences, SD^2. Then, as illustrated in Equation 8.3, we use the Spearman formula to determine a correlation coefficient:

rs = The Spearman rank-order correlation coefficient
D = The difference between the ranks on the two variables
n = The number of individuals or pairs of ranks

Calculation 8.2 (p. 176) shows how to use the formula to obtain a Spearman rank-order correlation coefficient.

Table 8.4

Twelfth Graders' GPAs and Combined SAT Scores

Student	GPA	Rank	SAT	Rank	D	D²
John Henry	3.97	1	1180	3	−1	1
Kate	3.85	2.5*	1130	6	−3.50	12.25
Wyatt	3.85	2.5*	1070	8	−5.50	30.25
Mattie	3.67	4	1180	3	1	1
Virgil	3.52	5	1140	5	0	0
Johnny	3.42	6	1290	1	5	25
Josephine	3.36	7	960	9	−2	4
Billy	3.24	8	1180	3	5	25
Jack	2.74	9	1110	7	2	4
Ike	2.56	10	720	10	0	0
						ΣD² = 102.50

When ties occur on *x* or *y*, assign the average of the rank involved to each score (e.g., $\frac{2+3}{2} = 2.50$).

Calculation 8.2

$$r_s = 1 - \frac{6\Sigma(D^2)}{n(n^2-1)}$$

Step 1

$$r_s = 1 - \frac{6\Sigma(105.5)}{10(10^2-1)}$$

Step 2

$$r_s = 1 - \frac{633}{990}$$

Step 3

$$r_s = 1 - 0.639 = 0.331$$

Calculation 8.2 shows that $r_s = 0.331$, which is interpreted as a weak positive relationship between GPA and combined SAT scores. Therefore, the example shows that GPA is not a very good predictor of SAT scores and that other predictors, such as motivaton and achievement test results, should be considered.

Check For Understanding

8.5. Which scale of measurement is required for a Spearman rank-order correlation?
   a. Ratio
   b. Interval
   c. Ordinal
   d. Nominal

8.6. Find the Spearman rank-order correlation coefficient between the two sets of scores.

   Scores

Student	Reading	IQ
A	98	145
B	91	135
C	88	125
D	85	120
E	80	110
F	75	105
G	70	95
H	74	95
I	65	90
J	60	80

8.7. Based on your calculated Spearman correlation coefficient from question 8.6, explain the strength of the relationship between reading scores and IQ scores (very weak, moderate, or strong, for instance) and determine the direction of the relationship (is it positive or negative)?

SUMMARY

The correlation coefficient helps us understand the relationship between two variables. The Pearson correlation coefficient, the most commonly used correlation statistic, is used with either an interval or a ratio scale of measurement. The Spearman rank-order correlation examines the relationship between two variables that are on an ordinal scale.

All coefficients of correlation range from −1.0 to +1.0, with the direction of the relationship shown by the sign (+ or −), and the strength of the relationship is based on its numerical value. For example, a correlation coefficient of zero indicates that there is no correlation between two variables, whereas a positive correlation shows that as the numerical values associated with one variable increase, the numerical values associated with the other variable increase as well. With a negative correlation, as one variable increases, the other variable decreases. A correlation statistic only shows the relative strength of a linear relationship between two variables. It does not imply a cause-and-effect relationship between two variables.

Classroom teachers use correlation to explore relationships between variables. For example, is intelligence related to academic achievement? The next two chapters cover validity and reliability, which deal with the concept of correlation. Therefore, to answer questions related to validity and reliability, an understanding of correlation is necessary.

CHAPTER REVIEW QUESTIONS

8.8. How is a Pearson correlation coefficient different from a Spearman correlation?

8.9. Explain how to calculate a Pearson correlation coefficient.

8.10. The numerical value for a correlation ranges between
a. 0.00 and +1.00.
b. –1.00 and 0.00.
c. –1.00 and +1.00.
d. none of the above.

8.11. What does the Pearson correlation coefficient measure?

8.12. A scatter plot shows a data set spread in a circular pattern. Which correlation coefficient would best describe the correlation for these data?
a. 0.00
b. 1.00
c. –0.50
d. all of the above

8.13. For a Pearson correlation coefficient of –1.90 between test X and test Y, the correlation indicates that
a. as scores on test X increase, scores on test Y decrease.
b. as scores on test X increase, scores on test Y increase.
c. as scores on test X decrease, scores on test Y decrease.
d. none of the above.

8.14. Calculate a Pearson correlation coefficient for the following sets of test scores.

Student	Test A	Test B
A	98	100
B	71	81
C	85	66
D	64	50

8.15. If the original data are measured on an ordinal scale of measurement, what type of correlation should you use?
a. Pearson
b. Spearman
c. Gaussian
d. all of the above

8.16. Calculate a Spearman correlation for the following sets of ranked test scores.

Student	SAT (combined)	GPA
A	1420	3.90
B	1100	3.00
C	1050	2.80
D	980	3.00
E	700	1.68

8.17. A researcher finds a high positive correlation between price of school lunches and achievement test scores. As a result, the researcher concludes that more expensive lunches cause test scores to increase. Do you agree or disagree with the researcher's conclusion? Please explain your decision.

ANSWERS: CHECK FOR UNDERSTANDING

8.1. c. + 0.15
8.2. + 0.733
8.3. A strong positive correlation
8.4. d. A high positive correlation
8.5. c. Ordinal
8.6. $r_s = 0.997$
8.7. There is a very strong positive correlation. This means as one score increases the other increases.

ANSWERS: CHAPTER REVIEW QUESTIONS

8.8. The Pearson and Spearman coefficients are mathematically identical. However, the Spearman rank coefficient is calculated from the ranks of each variable, not the actual values.

8.9. The Pearson correlation coefficient is calculated based on the following formula that uses your z-scores:

$$r = \frac{\sum z_x z_y}{n-1}$$

8.10. c. −1.00 and 1.00.

8.11. The purpose of the Pearson correlation coefficient is to indicate a linear relationship between two measurement variables. This means that if you have two sets of scores, you can determine if one score predicts another score.

8.12. a. 0.00

8.13. d. none of the above.

8.14. Correlation coefficient (r) = 0.7752

8.15. b. Spearman

8.16. Spearman r = 0.825 (corrected for ties)

8.17. Because correlation does not establish cause and effect, the research should not conclude causality.

INTERNET RESOURCES

This website of the University of Florida is linked to universities, statistical journals, mailing lists, and statistical software vendors. It is an excellent site for resource and reference information related to correlation. *www.stat.ufl.edu/vlib/statistics.html*

This Yahoo website offers links to some of the most popular statistics links on the internet. In addition, it provides links to statistics journals and online software. *http://dir.yahoo.com/science/mathematics/statistics*

This website of the American Statistical Association offers updated information about the field of statistics, professional journals, professional development courses, and careers. *www.amstat.org*

This website provides a free 30-day download of the Analyze-it software package. Analyze-it is an add-in for Microsoft Excel (for Windows) and is designed to calculate correlation coefficients, along with other statistical calculations that are covered in this text. *www.analyze-it.com*

REFERENCE

Christmann, E. P., and J. L. Badgett. 2000. A four-year analytic comparison of eleventh grade academic achievement in the Slippery Rock Area High School, and district pupil expenditures. Research report presented to The Slippery Rock School District's Board of Directors and The Pennsylvania Department of Education. (ERIC Document No. ED443824)

FURTHER READING

American Educational Research Association (AERA). 1999. *Standards for educational and psychological testing.* Washington, DC: American Educational Research Association.

Elmore, P. B., and P. L. Woehlke. 1997. *Basic statistics.* New York: Longman.

Gravetter, F. J., and L.B. Wallnau. 2002. *Essentials of statistics for the behavioral sciences.* New York: West.

Pearson, E. S. 1968. Some early correspondence between W. S. Gosset, R. A. Fischer, and Karl Pearson, with notes and comments. In *Studies in the history of statistics and probability,* eds. E. S. Pearson and M. G. Kendall. Vol. 1, pp. 405–417. London: Charles Griffin.

Raymond, J. C. 1999. *Statistical analysis in the behavioral sciences.* New York: McGraw-Hill.

Thorne, B. M., and J. M. Giesen. 2000. *Statistics for the behavioral sciences.* Mountain View, CA: Mayfield.

Zawojewski, J. S., and J. M. Shaughnessy. 2000. Mean and median: Are they really so easy? *Mathematics Teaching in the Middle School* 5 (7): 436–440.

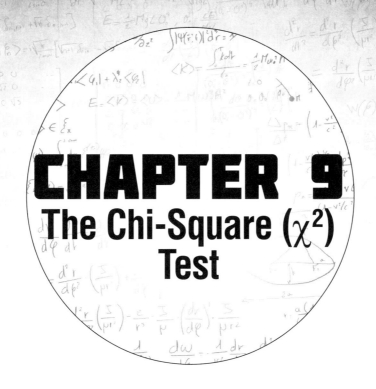

CHAPTER 9
The Chi-Square (χ^2) Test

OBJECTIVES

When you complete this chapter, you should be able to

1. demonstrate an understanding of the one-way Chi-Square (goodness of fit) and its relevance to classroom settings;
2. demonstrate an understanding of the two-way Chi-Square (independence) and its relevance to classroom settings;
3. compare the relationship between nonparametic procedures and parametric procedures;
4. calculate Chi-Squares from classroom and standardized test results; and
5. find, use, and interpret Chi-Square results.

Key Terms

When you complete this chapter, you should understand the following terms:
  nonparametric
  Chi-Square Goodness of Fit
  expected frequency
  phi coefficient
  Chi-Square of Independence

INTRODUCTION

The Chi-Square (χ^2) is a test statistic used to determine whether proportions in two or more categories differ significantly. The Chi-Square statistic is a nonparametric statistic, which means that data is taken from the nominal scale (e.g., number of males versus number of females) and does not assume a normal distribution of data. Nonetheless, nonparametric procedures will evaluate type-I errors by using an alpha level to determine a critical value from a χ^2 table and compare it to the statistical computation.

CHAPTER 9

It is important to know how to apply the χ^2 to research because data will not always be from interval and ratio scales of measurement. The χ^2 can analyze data from a nominal scale of measurement. Nominal scales clarify numbers into different (named) categories. In education, school districts categorize students based on grade levels (i.e., freshman, sophomores, juniors, and seniors). Hence, a teacher might find that of 20 randomly selected students, three are freshman, eight are sophomores, six are juniors, and four are seniors. Subsequently, the question is how can grouped categories represent, or be generalized to, a population of interest?

R. L. Plackett (1983) explains that Karl Pearson introduced the ***Chi-Square Test of Goodness of Fit*** in 1900. Pearson's test came out of his research on the "laws of chance" as related to experiments with the probability with tosses of a penny. As was the case in Pearson's experiment, the Chi-Square Test of Goodness of Fit is used when data consists of different categories with frequency data. Similar to correlation, we are exploring the relationship between the different categories and are determining whether there is a relationship between those categories.

CHI-SQUARE TEST OF GOODNESS OF FIT (χ^2)

Using data similar to Pearson's early experiment, we will examine the simplest case with $n = 100$ tosses of a penny.

Table 1 Penny Tossing Experiment

Heads	Tails
$f_o = 39$	$f_o = 61$

K = 2
N = total f_o = 100

In Table 1, each column displays the frequency within each category. More specifically, the individual columns are known as "cells," where the frequencies are numerical values referred to as observed frequencies (f_o). Hence, the f_o for heads is 39 and the f_o total for tails is 61. The combined total of observed frequencies is 100. It is also important to point out that the symbol (K) represents the number of categories. In Table 1, there are two categories (K = 2).

When conducting the Chi-Square Test of Goodness of Fit, I recommend adding percentages as descriptive statistics for the analysis. Table 2 shows this procedure:

Table 2 Descriptive Statistics for Penny Tossing Experiment

	Heads	Tails
Frequencies	$f_o = 39$	$f_o = 61$
Percentages	39%	61%

Once the descriptive statistics have been computed, we can begin the process of computing the Chi-Square Test of Goodness of Fit. The following procedure is the same seven-step process we used for all of the statistical tests in previous chapters.

Seven Steps for Hypothesis Testing

STEP 1: Identify the appropriate statistical test to be used with the given data for analysis.

STEP 2: Hypotheses
- H_o (Null Hypothesis): This hypothesis proposes that no statistical significance exists. The null hypothesis attempts to show that no variation exists between the means, or that a single variable is not different. When the null hypothesis is not rejected, it is presumed to be true until statistical evidence nullifies it for an alternative hypothesis.
- H_1 (Alternative Hypothesis): The alternative hypothesis is the hypothesis that is contrary to the null hypothesis. This hypothesis should correspond with the researcher's assumption and is used to discern the conclusion.

STEP 3: Alpha Level (α)
Alpha level is the probability of making a type-I error (rejecting the null hypothesis when the null hypothesis is true). The lower that this value is, the less chance that the researcher has of committing a type-I error. Also, an experiment should use a sample size large enough (i.e., $n = 20$) so that there is adequate statistical power. As a general rule, statistical analysis in education uses alpha levels of 0.05 or sometimes 0.01.

STEP 4: Critical Value (C.V.)
A number which causes rejection of the null hypothesis if a given test statistic is this number or more, and nonrejection of the null hypothesis if the test statistic is smaller than this number. The critical value can be found by analyzing a statistical table related to the appropriate statistical test.

STEP 5: Statistical Test
A statistical test provides a mechanism for making quantitative decisions. The intent is to determine whether there is enough evidence to "reject" the null hypothesis. Not rejecting may be a good result if we want to continue to act as if we "believe" the null hypothesis is true. Or it may be a disappointing result, possibly indicating we may not yet have enough data to "prove" something by rejecting the null hypothesis.

STEP 6: Reject or Do Not Reject the Null Hypothesis
By comparing the critical value to the calculated statistics, if the statistic exceeds the critical value, the null hypothesis is rejected.

> **STEP 7: Interpretation**
> Rejecting the null hypothesis may disclose that an independent variable/treatment did affect the dependent variable. However, not rejecting the null hypothesis may be a good result if we are looking for a result where no difference is desirable. Or it may be a disappointing result, possibly indicating we may not yet have enough data to "prove" something by rejecting the null hypothesis.

After understanding that the data must be analyzed by using a Chi-Square Test of Goodness of Fit, we can proceed with the hypothesis.

> **STEP 1: Identify the Statistical Test**
> Chi-Square Goodness of Fit

Most researchers test the null hypothesis (H_0), which is a statement that there is "no difference" among the frequencies within the categories when generalized to the population. Conversely, the alternative hypothesis (H_1) indicates that the frequencies are not equal. In this case the hypotheses are written:

> **STEP 2: Hypotheses**
> H_0: The frequencies in the population are equal.
> H_1: The frequencies in the population are different.

The third step is to set the alpha level (α). As explained earlier, the alpha level is used to determine the probability of committing a Type-I error. Different studies have advantages or disadvantages when using higher or lower α-levels. Hence, you have to be careful about interpreting the meaning of these terms. For example, *higher* α-levels can *increase* the chance of a Type-I error. Therefore, a lower α-level indicates that you are conducting a test that is more rigorous. Keep in mind that an α of 0.05 means that the researcher is taking a risk of being wrong 5 in 100 times by rejecting the null hypothesis (e.g., Committing a Type-I error is saying there's an effect when there really is not.). However, an alpha level (α) of 0.01 (compared with 0.05) means the researcher is being relatively cautious by minimizing the risk being wrong only 1 in 100 times to reject the null hypothesis. Based on the risk, the researcher decides the alpha level. In this case, the alpha level is 0.05.

> **STEP 3: Alpha Level (α)**
> $\alpha = 0.05$

The fourth step is to determine the critical value. The Critical Values Table (p. 204) for the Chi-Square is used to establish the critical value. First, you need to determine the Degrees of Freedom (df). This number is one less than the total number of categories (c),

(i.e., df = c – 1). Subsequently, there is only one degree of freedom, df = 2 – 1. Next we look at the Critical Values Table, which can be found at the end of the chapter (p. 204). The first column is under the alpha level of 0.05 and the first represents the value for a df = 1. You should find the Critical Value = 3.84.

> **STEP 4: Critical Value**
> C.V. = 3.84

The fifth step is to compute the Chi-Square Test of Goodness of Fit. The formula for the Chi-Square Goodness of Fit is:

$$\chi^2 = \sum (fo-fe)^2/fe$$

In this equation, *fo* represents the observed frequencies and *fe* represents the **expected frequencies**. The observed frequencies are tabulated by category, (heads, 39 and tails, 61, for example). However, to determine the *fe*, we start by organizing our data into categories. In this case, the categories are heads and tails and assume a 50/50 split of heads and tails. Therefore, we would expect an *fe* of heads to occur 50 times and an *fe* of tails to occur 50 times (Keep in mind that the assumption for *fe* changes for different scenarios and we might have a situation where our hypothesis is more complex (e.g., we could have an *fe* of 60/40 or another combination. If this is the case, we adjust our expected frequencies accordingly. However, if the *fe* for the categories is an unknown assumption, it is safe to assume that the fe should be equally divided among categories).

To calculate the Chi-Square Goodness of Fit statistic, we must plug in the observed and expected frequencies into the given equation.

> **STEP 5: Statistical Test (Chi-Square Goodness of Fit)**
> $$\chi^2 = \sum (fo-fe)^2/fe$$
> Step #1
> $$\chi^2 = (39-50)^2/50 + (61-50)^2/50$$
> Step #2
> $$\chi^2 = 2.42 + 2.42$$
> Step #3
> $$\chi^2 = 4.84$$

The calculation for the Chi-Square is $\chi^2 = 4.84$, which measures the difference between *fo* and *fe* in the categories. Notice that since each difference is a squared value, the χ^2 can never be a negative number.

The sixth step is to reject or to not reject the null hypothesis. This step takes into consideration a comparison of the critical value to the calculated Chi-Square statistic, which

compares the χ^2 of 4.84 to the C.V. of 3.84. Subsequently, since the Chi-Square statistic exceeds the critical value of 3.84, the null hypothesis is rejected.

> **STEP 6: Reject or Do Not Reject the Null Hypothesis**
> Since the χ^2 of 4.84 is larger than C.V. of 3.84, we reject the null hypothesis.

The seventh step is to interpret our findings. Since we rejected our null hypothesis, we conclude that the observed frequencies are significantly different from what we would expect with the coin tosses, as they would occur in a population. Therefore, questions about the consistency of the coin toss technique and/or whether or not the coin is a "fair" coin should be investigated.

> **STEP 7: Interpretation**
> Rejecting the null hypothesis may disclose that an independent variable/treatment did affect the dependent variable.
> There is a significant categorical difference between the frequency of heads and the frequency of tails when tossing a coin. In this case, when the coin is tossed, the frequency of tails is significantly higher than the frequency of heads.

CHI-SQUARE OF INDEPENDENCE (χ^2)

Since the Chi-Square (χ^2) is a method to answer questions from nominal data occurring as frequencies, the data can come in increased complexity when additional variables. However, we will introduce the Chi-Square of Independence (χ^2) in its most simplistic form (i.e., a two-way Chi-Square or a 2 × 2). Keep in mind that Chi-Squares of Independence can come as 2 × 3, 4 × 3, and so on. However, the procedures for computing the Chi-Square of Independence are the same for any of the scenarios.

Here is a study that could be analyzed with a two-way Chi-Square. A test of whether or not there is a significant difference between girls or boys who prefer listening to their iPod or reading a book for leisure.

	iPod	Book
Girls	fo = 17	fo = 35
Boys	fo = 171	fo =15

The first step is determined by examining what is known. Since there are two variables with two levels for each category and we have nominal scale data that is categorized by frequency, the data must be analyzed by using a Chi-Square of Independence.

> **STEP 1: Identify the Statistical Test**
> Chi-Square of Independence

The second step is the hypotheses. In this case, the hypotheses are written the same as the previous example of the Chi-Square Goodness of Fit. Again, the null hypothesis (H_o) is a statement that there is "no difference" among the frequencies within the categories when generalized to the population, and the alternative hypothesis (H_1) indicates that the frequencies are not equal. In this case the hypotheses are written:

> **STEP 2: Hypotheses**
> H_o: The frequencies in the population are equal.
> H_1: The frequencies in the population are different.

The third step is to set the alpha level (α). For this example, the researcher decides the alpha level α be set at 0.01.

> **STEP 3: Alpha Level (α)**
> $\alpha = 0.01$

The fourth step is to determine the critical value. Again, the Critical Values Table for the Chi-Square is used to establish the critical value. First, you need to determine the degrees of freedom (df). For the Chi-Square of Independence, the degree of freedom is calculated by using the following df = (R − 1)(C − 1), where R = the number of rows and C = the number of columns. Since there are two rows and two columns for this scenario, the df = (2 − 1)(2 − 1) = 1. Once we determine the df, we look at the critical values table at the end of the chapter. The second column is under the alpha level of 0.01 and the first represents the value for a df = 1. You should find the Critical Value = 6.64.

> **STEP 4: Critical Value**
> C.V. = 6.64

The fifth step is to compute the Chi Square of Independence. The formula for the Chi of Independence is:

$$\chi^2 = \Sigma(fo-fe)^2/fe$$

However, the Chi-Square of Independence requires another step in the procedure to compute the expected frequencies (fe). The formula to calculate the *fe* is:

$$fe = (fc)(fr)/n$$

In this equation, *fc* is the frequency in the columns and *fr* is the frequency in the rows. Therefore, to determine the expected frequencies, we must first add the observed frequencies in the columns and the observed frequencies in the rows:

	iPod	Book	
Girls	fo = 17	fo = 35	fr = 17 + 35 = 52
Boys	fo = 171	fo =15	fr = 171 + 15 = 186
	fc = 17 + 171 = 188	fc = 35 + 15 = 50	

The sample size, *n*, is computed by adding all of the observed frequencies within the four cells. Therefore, n = 17 + 35 + 171 + 15 = 238. Next, we use the expected frequency formula to compute the expected frequency for each cell:

	iPod	Book
Girls	188 × 52/238 fe = 41.08	50 × 52/238 fe = 10.92
Boys	188 × 186/238 fe = 146.92	50 × 186/238 fe = 39.08

Once we have calculated the expected frequencies within the cells, we can use the formula for the Chi Square of Independence:

> **STEP 5: Statistical Test (Independence)**
> $\chi^2 = \Sigma(fo-fe)^2/fe$
> $\chi^2 = (17 - 41.08)^2/ 41.08 + (35 - 10.92)^2/ 10.92 +(171 - 146.92)^2/ 146.92 + (15 - 39.08)^2/ 39.08$
> $\chi^2 = 14.12 + 53.10 + 3.95 + 14.84 = 86.01$

The sixth step is to reject or to not reject the null hypothesis. This step takes into consideration a comparison of the critical value to the calculated Chi-Square statistic, which compares the χ^2 of 86.01 to the C.V. of 6.64. Subsequently, since the Chi-Square statistic exceeds the critical value of 6.64, the null hypothesis is rejected.

> **STEP 6: Reject or Do Not Reject the Null Hypothesis**
> Since the χ^2 of 86.01 is larger than the C.V. of 6.64, we reject the null hypothesis.

Is There a Statistical Relationship?

Once we have computed the Chi-Square of Independence statistic, we can use a procedure to determine whether a significant correlation exists. A procedure that is used with a 2 × 2 contingency table is known as the **phi coefficient** (ϕ). As is the case with correlation, the value of a phi coefficient ranges between 0 and 1. The interpretation of the phi coefficient is the same as that used for the Pearson coefficient (see Chapter 8 on correlation). If a contingency table is larger than 2 × 2, a contingency coefficient (C) is used.

Phi Coefficient Equation

$$\sigma = \sqrt{\frac{\chi^2}{n}}$$

Contingency Coefficient Equation

$$C = \sqrt{\frac{\chi^2}{N + \chi^2}}$$

where
χ^2 = the chi-square test statistic given above
N = the total sample size

Phi Coefficient Computation

$0.601 = \sqrt{86.01/238}$

As with correlation, the larger the value for the phi coefficient (ϕ), the stronger the relationship and the closer the data come to being dependent. If you refer to Table 8.1 Interpretations of Correlations, you can see that this is a strong positive correlation, meaning that there is a relationship between boys who prefer listening to their iPod.

The seventh step is to interpret our findings. Since we rejected our null hypothesis, we conclude that the observed frequencies are significantly different from what we would expect between girls and boys who prefer listening to their iPod or reading a book for leisure. In this case, the most significant frequency is the number of boys who prefer listening to their iPod.

STEP 7: Interpretation

Rejecting the null hypothesis may disclose that an independent variable/treatment did affect the dependent variable.

CALCULATOR EXPLORATION

Using the TI-83 or TI-84, you can easily compute a Chi-Square of Independence by following these keystrokes.

Problem

	Success	Failure
Method A	25	75
Method B	40	160

We need the chi-square critical value associated with this data. As shown in our example, the degrees of freedom for this problem is df = (2 – 1)(2 – 1) = 1.

Calculating Chi-Square Statistic

Press [**2nd MATRIX**].
Select [**EDIT -> 1: A**].
Copy the following data by typing in each number and then pressing [**ENTER**].

Now Press [**STAT**]. Under the [**TESTS**] submenu scroll down and select [**C: X² -Test…**]. Press [**ENTER**].

Move cursor down to [**DRAW**] and press [**ENTER**].

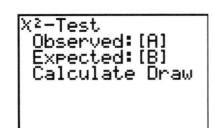
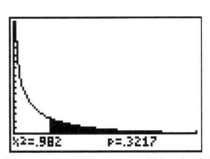

The Chi-Square statistic is 0.982. Since 0.982 < 3.841... and p-value is p= 0.3217 which is greater than 0.05, we do not reject the null hypothesis.

SUMMARY

Although less powerful than parametric statistics, nonparametric procedures are used when the data do not meet the assumptions of parametric procedures (e.g., parametric statistical tests assume that your data are normally distributed and use statistical tests such as the t-test). The Chi-Square statistic is used with data presented in a nominal scale of measurement with more than one category as frequencies.

The Chi-Square Test of Goodness of Fit is a statistical test that the fit of a set of categorical observations. Hence, the Chi-Square Goodness of Fit typically summarizes the discrepancy between observed values and the values expected under the case in question.

The Chi-Square of Independence is a statistical test that is applied when you have two categorical variables from a single population. It is used to determine whether there is a significant association between the two variables. For example, in a survey, respondents might be classified by gender (boys or girls) and hand preference (right handed or left handed). We could use aChi-Square Test for Independence to determine whether gender is related to hand preference. Once the Chi-Square of Independence is calculated, the phi coefficient (ϕ) is used to estimate the strength of the relationship.

CHAPTER REVIEW QUESTIONS

1. Using the Seven Steps for Hypothesis Testing, analyze the given survey question's results at the 0.05 level.

Survey Question
How would you rate the quality of the overall for scientific inquiry program?

Very High 5	High 4	Average 3	Low 2	Very Low 1

Survey Question Results

Table 1

	Descriptive statistics				
	5	4	3	2	1
Frequencies	37	65	20	0	1

ANSWER FOR QUESTION 1

Seven Steps for Hypothesis Testing

STEP 1: Chi-Square Test of Goodness of Fit
STEP 2: Hypotheses H_o: The frequencies in the population are equal. H_1: The frequencies in the population are different.
STEP 3: Alpha Level (α)   Alpha level is 0.05.
STEP 4: Critical Value (C.V.)   C.V. = 9.49
STEP 5: Statistical Test $\chi^2 = 123.656$
STEP 6: Reject Null Hypothesis
STEP 7: Interpretation   *Summary*   The survey respondents for the science inquiry program for in-service teachers indicated at a statistically significant level, χ^2 (4, n = 122) = 123.656, p < 0.05, that the quality of the overall program is "High."

2. Using the Seven Steps for Hypothesis Testing, analyze the given survey question's results at the 0.05 level.

Survey Question

The science inquiry program was focused on a timely topic.

Strongly Agree	Agree	Undecided	Disagree	Strongly Disagree
5	4	3	2	1

Survey Question Results

Table 2

	Descriptive statistics				
	5	4	3	2	1
Frequencies	70	52	1	0	1

ANSWER FOR QUESTION 2

Seven Steps for Hypothesis Testing

STEP 1: Chi-Square Test of Goodness of Fit
STEP 2: Hypotheses H_o: The frequencies in the population are equal. H_1: The frequencies in the population are different.
STEP 3: Alpha Level (α)   Alpha level is 0.05.
STEP 4: Critical Value (C.V.)   C.V. = 9.49
STEP 5: Statistical Test $\alpha^2 = 182.694$
STEP 6: Reject Null Hypothesis
STEP 7: Interpretation

Summary

The survey respondents for the scientific inquiry program indicated at a statistically significant level, χ^2 (4, n = 124) = 182.694, p < 0.05, that they "Strongly Agree" the program was focused on a timely topic.

3. Using the Seven Steps for Hypothesis Testing, analyze the given survey question's results at the 0.01 level.

Survey Question

How valuable was the scientific inquiry program in helping you meet your goals?

Very Valuable	Somewhat Valuable	Not Very Valuable	Not at All Valuable
4	3	2	1

Survey Question Results

Table 3

	Descriptive statistics			
	4	3	2	1
Frequencies	83	36	1	0

ANSWER FOR QUESTION 3

Seven Steps for Hypothesis Testing

STEP 1: Chi-Square Test of Goodness of Fit
STEP 2: Hypotheses H_o: The frequencies in the population are equal. H_1: The frequencies in the population are different.
STEP 3: Alpha Level (α)   Alpha level is 0.01.
STEP 4: Critical Value (C.V.)   C.V. = 11.34
STEP 5: Statistical Test χ^2 =152.867
STEP 6: Reject Null Hypothesis
STEP 7: Interpretation   *Summary*   The survey respondents for the scientific inquiry program indicated at a statistically significant level, χ^2 (3, n = 120) = 152.867, p < 0.01, that the program was "Very Valuable" in helping to meet goals.

4. Using the Seven Steps for Hypothesis Testing, analyze the given survey question's results at the 0.05 level.

Survey Question

Course objectives/goals for the scientific inquiry program were clear throughout the sessions.

Strongly Agree	Agree	Undecided	Disagree	Strongly Disagree
5	4	3	2	1

Survey Question Results

Table 4

Descriptive statistics

	5	4	3	2	1
Frequencies	33	76	11	4	0

ANSWER FOR QUESTION 4

Seven Steps for Hypothesis Testing

STEP 1: Chi-Square Test of Goodness of Fit
STEP 2: Hypotheses H_0: The frequencies in the population are equal. H_1: The frequencies in the population are different.
STEP 3: Alpha Level (α)   Alpha level is 0.05.
STEP 4: Critical Value (C.V.)   C.V. = 9.49
STEP 5: Statistical Test $\chi^2 = 158.339$
STEP 6: Reject Null Hypothesis
STEP 7: Interpretation   *Summary*   The survey respondents for the scientific inquiry program indicated at a statistically significant level, χ^2 (4, n = 124) = 158.339, p < 0.05, that they "Agree" that course objectives/goals were clear throughout the sessions.

5. Using the Seven Steps for Hypothesis Testing, analyze the given survey question's results at the 0.05 level.

Survey Question

The scientific inquiry sessions had a clear purpose.

Strongly Agree	Agree	Undecided	Disagree	Strongly Disagree
5	4	3	2	1

Survey Question Results

Table 5

	Descriptive statistics				
	5	4	3	2	1
Frequencies	48	63	11	2	0

ANSWER FOR QUESTION 5

Seven Steps for Hypothesis Testing

STEP 1: Chi-Square Test of Goodness of Fit
STEP 2: Hypotheses H_o: The frequencies in the population are equal. H_1: The frequencies in the population are different.
STEP 3: Alpha Level (α)   Alpha level is 0.05.
STEP 4: Critical Value (C.V.)   C.V. = 9.49
STEP 5: Statistical Test $\chi^2 = 133.984$
STEP 6: Reject Null Hypothesis
STEP 7: Interpretation   *Summary*   The survey respondents for the advocate program indicated at a statistically significant level, χ^2 (4, n = 124) = 133.984, p < 0.05, that they "Agree" that the sessions had a clear purpose.

6. To better determine what technologies subject area teachers in a school district desire, a survey by the administration of two different subject areas was administered to mathematics teachers and science teachers in a school district. The two groups were asked for their preference in using either graphing calculators or personal digital assistants (PDAs). Determine the appropriate statistical test at the 0.05 alpha level.

	Calculators	PDAs
Science Teachers	fo = 21	fo = 19
Mathematics Teachers	fo = 30	fo = 10

ANSWER FOR QUESTION 6

Seven Steps for Hypothesis Testing

STEP 1: Chi-Square Test of Goodness of Fit
STEP 2: Hypotheses H_0: The frequencies in the population are equal. H_1: The frequencies in the population are different.
STEP 3: Alpha Level (α)   Alpha level is 0.05.
STEP 4: Critical Value (C.V.)   C.V. = 3.84
STEP 5: Statistical Test $\chi^2 = 4.381$ Phi Coefficient Computation $0.234 = \sqrt{4.381/80}$
STEP 6: Reject Null Hypothesis
STEP 7: Interpretation

Summary

The survey respondents for technologies indicated at a statistically significant level, χ^2 (1, n = 80) = 4.381, p <.05, that graphing calculators are the preference. The Phi Coefficient of 0.234 shows a weak relationship.

7. A survey of two groups of teachers, one group from an urban school and one group from a rural school was asked if they prefer teaching students for 45-minute classes or 90-minute (block schedule) classes. Determine the appropriate statistical test at the 0.05 alpha level.

	45-Minute	90-Minute
Urban Schools	fo = 35	fo = 15
Rural School	fo = 20	fo = 30

ANSWER FOR QUESTION 7

Seven Steps for Hypothesis Testing

STEP 1: Chi-Square Test of Goodness of Fit
STEP 2: Hypotheses H_o: The frequencies in the population are equal. H_1: The frequencies in the population are different.
STEP 3: Alpha Level (α)   Alpha level is 0.05.
STEP 4: Critical Value (C.V.)   C.V. = 3.84
STEP 5: Statistical Test $\chi^2 = 9.091$ Phi Coefficient Computation $0.302 = \sqrt{9.091/100}$
STEP 6: Reject Null Hypothesis
STEP 7: Interpretation

Summary

The survey respondents indicated at a statistically significant level, χ^2 (1, n = 100) = 9.091, $p < 0.05$, that 45-minute classes are preferred by urban teachers and 90-minute classes are preferred by rural teachers. The phi coefficient of 0.302 shows a weak relationship.

8. A survey of two groups of science teachers, physics and biology, was asked if they prefer having an extended laboratory period. Determine the appropriate statistical test at the 0.01 alpha level.

	Reg. Lab	Extended Lab
<u>Physics</u>	fo = 25	fo = 10
<u>Biology</u>	fo = 30	fo = 5

ANSWER FOR QUESTION 8

Seven Steps for Hypothesis Testing

STEP 1: Chi-Square Test of Goodness of Fit
STEP 2: Hypotheses H_o: The frequencies in the population are equal. H_1: The frequencies in the population are different.
STEP 3: Alpha Level (α)   Alpha level is 0.01.
STEP 4: Critical Value (C.V.)   C.V. = 6.64
STEP 5: Statistical Test $\chi^2 = 2.121$ Phi Coefficient Computation $0.174 = \sqrt{2.121/70}$
STEP 6: Do Not Reject Null Hypothesis
STEP 7: Interpretation
Summary   The survey respondents did not indicate at a statistically significant level, χ^2 (1, n = 70) = 2.121, p < 0.01, that there is a preference for extending the laboratories. The phi coefficient of 0.174 shows a very weak relationship.

9. A survey of science supervisors in Virginia and Pennsylvania were asked if they prefer doing traditional frog dissections or computer simulated frog dissections. Determine the appropriate statistical test at the 0.05 alpha level.

	Traditional	Computer Simulated
Virginia	fo = 121	fo = 79
Pennsylvania	fo = 130	fo = 70

ANSWER FOR QUESTION 9

Seven Steps for Hypothesis Testing

STEP 1: Chi-Square Test of Goodness of Fit
STEP 2: Hypotheses H_o: The frequencies in the population are equal. H_1: The frequencies in the population are different.
STEP 3: Alpha Level (α)   Apha level is 0.05.
STEP 4: Critical Value (C.V.)   C.V. = 3.84
STEP 5: Statistical Test $\chi^2 = 0.866$ Phi Coefficient Computation $0.047 = \sqrt{.866/400}$
STEP 6: Do Not Reject Null Hypothesis
STEP 7: Interpretation   *Summary*   The survey respondents did not indicate at a statistically significant level, χ^2 (1, n = 400) = 0.866, p < 0.05, that there is a preference for extending the laboratories. The Phi Coefficient of 0.047 shows a very weak relationship.

10. A survey of middle-level science students from Europe and the United States were asked if they prefer doing labs or participating in lectures. Determine the appropriate statistical test at the 0.05 alpha level.

	Labs	Lectures
U.S.	fo = 1506	fo = 494
Europe	fo = 1320	fo = 680

ANSWER FOR QUESTION 10

Seven Steps for Hypothesis Testing

STEP 1: Chi-Square Test of Goodness of Fit
STEP 2: Hypotheses H_o: The frequencies in the population are equal. H_1: The frequencies in the population are different.
STEP 3: Alpha Level (α)   Alpha level is 0.05.
STEP 4: Critical Value (C.V.)   C.V. = 3.84
STEP 5: Statistical Test $\chi^2 = 41.711$ Phi Coefficient Computation $0.047 = \sqrt{41.711/4000}$
STEP 6: Do Not Reject Null Hypothesis
STEP 7: Interpretation   *Summary*   The survey respondents indicated at a statistically significant level, χ^2 (1, n = 4000) = 41.711, $p < 0.05$, that there is a preference for laboratories over lectures in the United States and Europe. The Phi Coefficient of 0.102 shows a very weak relationship.

REFERENCE

Plackett, R. L. 1983. Karl Pearson and the chi-squared test. *International Statistical Review / Revue Internationale De Statistique* 51: 1, 59–72.

FURTHER READING

Baum, S., R. K. Gable, and K. List. 1987. *Chi square, pie charts, and me.* Monroe, NY: Trillium Press.

Hewett, J. E., and R. K. Tsutakawa. 1971. *Two-stage chi square goodness of fit test.* Columbia: University of Missouri--Columbia.

Molinari, L., and University of California at Berkeley Statistical Lab. 1975. *Distribution of the chi-square test in nonstandard situations.* Ft. Belvoir, VA: Defense Technical Information Center.

CHAPTER 9

Critical Values Table for the Chi-Square

df	P = 0.05	P = 0.01
1	3.84	6.64
2	5.99	9.21
3	7.82	11.35
4	9.49	13.28
5	11.07	15.09
6	12.59	16.81
7	14.07	18.48
8	15.51	20.09
9	16.92	21.67
10	18.31	23.21
11	19.68	24.73
12	21.03	26.22
13	22.36	27.69
14	23.69	29.14
15	25.00	30.58
16	26.30	32.00
17	27.59	33.41
18	28.87	34.81
19	30.14	36.19
20	31.41	37.57
21	32.67	38.93
22	33.92	40.29
23	35.17	41.64
24	36.42	42.98
25	37.65	44.31
26	38.89	45.64
27	40.11	46.96
28	41.34	48.28
29	42.56	49.59
30	43.77	50.89
31	44.99	52.19
32	46.19	53.49
33	47.40	54.78
34	48.60	56.06
35	49.80	57.34
36	51.00	58.62
37	52.19	59.89
38	53.38	61.16
39	54.57	62.43
40	55.76	63.69
41	56.94	64.95
42	58.12	66.21
43	59.30	67.46
44	60.48	68.71
45	61.66	69.96
46	62.83	71.20
47	64.00	72.44
48	65.17	73.68
49	66.34	74.92
50	67.51	76.15
51	68.67	77.39
52	69.83	78.62
53	70.99	79.84
54	72.15	81.07
55	73.31	82.29
56	74.47	83.52

57	75.62	84.73		89	112.02	122.94
58	76.78	85.95		90	113.15	124.12
59	77.93	87.17		91	114.27	125.29
60	79.08	88.38		92	115.39	126.46
61	80.23	89.59		93	116.51	127.63
62	81.38	90.80		94	117.63	128.80
63	82.53	92.01		95	118.75	129.97
64	83.68	93.22		96	119.87	131.14
65	84.82	94.42		97	120.99	132.31
66	85.97	95.63		98	122.11	133.47
67	87.11	96.83		99	123.23	134.64
68	88.25	98.03		100	124.34	135.81
69	89.39	99.23				
70	90.53	100.42				
71	91.67	101.62				
72	92.81	102.82				
73	93.95	104.01				
74	95.08	105.20				
75	96.22	106.39				
76	97.35	107.58				
77	98.49	108.77				
78	99.62	109.96				
79	100.75	111.15				
80	101.88	112.33				
81	103.01	113.51				
82	104.14	114.70				
83	105.27	115.88				
84	106.40	117.06				
85	107.52	118.24				
86	108.65	119.41				
87	109.77	120.59				
88	110.90	121.77				

Appendix

AREAS OF THE STANDARD NORMAL DISTRIBUTION

z	.00	.01	.02	.03	.04	.05	.06	.07	.08	.09
0.0	.0000	.0040	.0080	.0120	.0160	.0199	.0239	.0279	.0319	.0359
0.1	.0398	.0438	.0478	.0517	.0557	.0596	.0636	.0675	.0714	.0753
0.2	.0793	.0832	.0871	.0910	.0948	.0987	.1026	.1064	.1103	.1141
0.3	.1179	.1217	.1255	.1293	.1331	.1368	.1406	.1443	.1480	.1517
0.4	.1554	.1591	.1628	.1664	.1700	.1736	.1772	.1808	.1844	.1879
0.5	.1915	.1950	.1985	.2019	.2054	.2088	.2123	.2157	.2190	.2224
0.6	.2257	.2291	.2324	.2357	.2389	.2422	.2454	.2486	.2517	.2549
0.7	.2580	.2611	.2642	.2673	.2704	.2734	.2764	.2794	.2823	.2852
0.8	.2881	.2910	.2939	.2967	.2995	.3023	.3051	.3078	.3106	.3133
0.9	.3159	.3186	.3212	.3238	.3264	.3289	.3315	.3340	.3365	.3389
1.0	.3413	.3438	.3461	.3485	.3508	.3531	.3554	.3577	.3599	.3621
1.1	.3643	.3665	.3686	.3708	.3729	.3749	.3770	.3790	.3810	.3830
1.2	.3849	.3869	.3888	.3907	.3925	.3944	.3962	.3980	.3997	.4015
1.3	.4032	.4049	.4066	.4082	.4099	.4115	.4131	.4147	.4162	.4177
1.4	.4192	.4207	.4222	.4236	.4251	.4265	.4279	.4292	.4306	.4319
1.5	.4332	.4345	.4357	.4370	.4382	.4394	.4406	.4418	.4429	.4441
1.6	.4452	.4463	.4474	.4484	.4495	.4505	.4515	.4525	.4535	.4545
1.7	.4554	.4564	.4573	.4582	.4591	.4599	.4608	.4616	.4625	.4633
1.8	.4641	.4649	.4656	.4664	.4671	.4678	.4686	.4693	.4699	.4706
1.9	.4713	.4719	.4726	.4732	.4738	.4744	.4750	.4756	.4761	.4767
2.0	.4772	.4778	.4783	.4788	.4793	.4798	.4803	.4808	.4812	.4817
2.1	.4821	.4826	.4830	.4834	.4838	.4842	.4846	.4850	.4854	.4857
2.2	.4861	.4864	.4868	.4871	.4875	.4878	.4881	.4884	.4887	.4890
2.3	.4893	.4896	.4898	.4901	.4904	.4906	.4909	.4911	.4913	.4916
2.4	.4918	.4920	.4922	.4925	.4927	.4929	.4931	.4932	.4934	.4936
2.5	.4938	.4940	.4941	.4943	.4945	.4946	.4948	.4949	.4951	.4952
2.6	.4953	.4955	.4956	.4957	.4959	.4960	.4961	.4962	.4963	.4964
2.7	.4965	.4966	.4967	.4968	.4969	.4970	.4971	.4972	.4973	.4974
2.8	.4974	.4975	.4976	.4977	.4977	.4978	.4979	.4979	.4980	.4981
2.9	.4981	.4982	.4982	.4983	.4984	.4984	.4985	.4985	.4986	.4986
3.0	.4987	.4987	.4987	.4988	.4988	.4989	.4989	.4989	.4990	.4990

Index

*Page numbers printed in **boldface** type refer to figures, tables, or equations.*

A

Achenwall, Gottfried, 2
Alcuin of York, 2
Alpha level (α), 70, 72, 80, 95, 126, 127, 142, 151, 185
    for analysis of variance, 143–144
    for chi-square of independence, 189
    for chi-square test of goodness of fit, 186
    critical value and, 70, 74–75, 77, 81
    for dependent t-test
        one-tailed test, 120
        two-tailed test, 116
    for independent t-test
        one-tailed test, 105–106
        two-tailed test, 110, 112
    for one-sample t-test
        one-tailed test, 100
        two-tailed test, 96–97
    for one-sample z-test
        one-tailed -test, 77
        two-tailed test, 74
Alternative hypothesis (H_1), 72, 73, 76, 80, 95, 126, 142, 151, 185
    for analysis of variance, 143
    for chi-square of independence, 189
    for chi-square test of goodness of fit, 186
    for dependent t-test
        one-tailed test, 120
        two-tailed test, 116
    for independent t-test
        one-tailed test, 105
        two-tailed test, 110
    for one-sample t-test

INDEX

one-tailed test, 99
two-tailed test, 96
for one-sample z-test
one-tailed test, 76–77
two-tailed test, 73
An Argument for Divine Providence, Taken From the Constant Regularity Observed in the Births of Both Sexes (Arbuthnot), 70
Analysis of variance (ANOVA), 4, 6, 141–162
one-way, 141–146
computing with graphing calculator, 149–150
formula for, 144
steps in calculation of, 142–146
use of, 141–142, 150–151
post-hoc tests and, 146–149
Scheffe's method for calculation of, 146–149
review questions on, 152–162
Arbuthnot, John, 70
Archimedes, 2, 5
Areas of the standard normal distribution, 75, 207

B

Bar graph
histogram, 14, **16,** 19
simple, 14, **16,** 19
Bell (normal) curve, 16–17, **17,** 53, 55. See also Normal distributions
mean of scores on, 30, **30**
standard deviation and, 35, **36**
Bimodal distribution, 32
Biometrika (Pearson), 4, 6

C

Calculator explorations
computing chi-square of independence, 192–193
computing dependent t-test, 123–125
computing independent t-test, 113–115
computing mean and standard deviation, 45–46

computing one-sample t-test, 102–103
computing one-way analysis of variance, 149–150
computing Pearson correlation coefficient, 168–170
computing percentage of students in normal distribution with IQ between 85 and 130, 64–65
computing z-test, 79–80
sorting numbers, 23–25
Cartesian grid, 13, **15**
Cause-and-effect and correlation, 170, 177
Celsius temperature scale, 10
Central tendency measures, 27–33, 40
definitions of, 29
mean, 30–31
media examples of, 29
median, 31–32
mode, 32–33
in normal distribution, 29, **30**
practical uses for teachers, 28, **28,** 29
review questions on, 41–45
symbols used in computation of, **30**
variation and, 28, 34
Charlemagne, 2
Chauncey, Henry, 57
Chi-square (χ^2) test, 183–205
application of, 184, 188, 193
chi-square of independence, 188–193
application of, 188, 193
computation of, 189–191
computing with graphing calculator, 192–193
formula for, 189–190
observed frequencies and expected frequencies for, 190
phi coefficient to determine statistical relationship, 190
definition of, 183
of goodness of fit, 184–188, 193
computation of, 185–188
descriptive statistics for, 184
formula for, 187
observed frequency and expected frequency for, 184, 187
Class ranks, 51–52
Conant, James Bryant, 57
Continuous variables, 11–12, **12,** 19

INDEX

Coordinate axes, 13, **13**
Correlation, 165–180
    cause-and-effect and, 170, 177
    definition of correlation coefficient, 166
    interpretation of, 166, **166**
    numerical values of, 166, 175, 177
    origin of term for, 3
    Pearson correlation coefficient, 3, 165–170
    review questions on, 178–180
    scatter plots of, 170–173, **171**, 177
        high positive correlation, 172, **172**
        low negative correlation, 173, **173**
        no correlation, 173, **173**
        perfect negative correlation, 172, **172**
        perfect positive correlation, 171, **171**
    between smoking and lung cancer, 3, 5
    Spearman rank-order correlation coefficient, 175–177
    uses of, 165–166, 177–178
Criterion-referenced data, 35, 49
Critical value (C.V.), 70, 72, 80, 91, 126, 127, 143, 151, 185
    alpha level and, 70, 74–75, 77, 81
    for analysis of variance, 144
    for chi-square test, 186–187, 189, **204–205**
    for dependent t-test
        one-tailed test, 121
        two-tailed test, 116–117
    for independent t-test
        one-tailed test, 105–106
        two-tailed test, 110, 112
    for one-sample t-test
        one-tailed test, 100
        two-tailed test, 97–98
    for one-sample z-test
        one-tailed test, 77
        two-tailed test, 74–75

D
Data
    criterion-referenced, 35, 49
    gathering of, 2, 6
    norm-referenced, 35, 50, 53, 55, 64
Data organization, 12–19

frequency distributions, 12, **12, 13**
frequency graphs, **13**, 13–16, **15–16**
frequency table, 13, **14**
normal distribution, 16–17, **17**
review questions on, 20–23
rounding numbers, 19
skewed distributions, 17–19, **19**
Degrees of freedom (df), 4, 186, 189, 192
    analysis of variance and, 144, **163–164**
    chi-square test of goodness of fit and, 186–187
    t-test and, 94, 97, 104, 106, 110, 116, 121
Dependent t-test, 115–125
    computing with graphing calculator, 123–125
    example of one-tailed test, 119–123
    example of two-tailed test, 115–119
    use of, 115
Descartes, Rene, 13
Discrete variables, 11, 12, 19
Distance measurements, 11

E
Educational Testing Service (ETS), 55
Einstein, Albert, 3
Euclid, 2
Expected frequency, 187, 190

F
Fahrenheit, Gabriel, 11
Fahrenheit temperature scale, 10, 11
False positive results. *See* Type-I error
Fingerprint analysis, 3
Fisher, Sir Ronald Aylmer, 4–5, 70, 81, 94, 127, 141
F-ratio, 144–148
Frequency distributions, 12, **12, 13**
Frequency graphs, **13**, 13–14, **15–16**
Frequency polygon, 13–14, **15**, 19
Frequency table, 13, **14**
F-Table, 144, **163–164**

INDEX

G

Galton, Francis, 3, 53
Gender ratio at birth, 70
Gosset, William Sealy, 4, 93–94
Gould, Steven Jay, 18
Grade point averages (GPAs), 51, **51**
Graphing calculators, 5. *See also* Calculator explorations
Groups, homogeneous and heterogeneous, 28

H

Hall, G. Stanley, 53
Hellenistic Age, 2, 6
Hernstein, Richard, 18
Heterogeneous groups, 28
Histogram, 14, **16,** 19
History of statistics, 2–7
    early statisticians, 2–5
    review questions on, 6–7
Homogeneous groups, 28
Hypothesis testing
    for analysis of variance, 141–162
    for chi-square test of goodness of fit, 185–187
    seven steps for, 72–73, 80–81, 95–96, 126–127, 142–143, 151–152, 185–186
    for t-test, 93–116
    for z-test, 69–91

I

Independent t-test, 103–115, 126
    computing with graphing calculator, 113–115
    example of one-tailed test, 105–109
    example of two-tailed test, 109–113
    formulas for, 104
    use of, 104
Inferential statistics, 3, 74, 80, 81, 127
Intelligence (IQ)
    as discrete variable, 12
    interpreting scores on tests of, 54
        Iowa Test of Basic Skills, 62
    measurement of, 11
    normal distribution of, 17
    one-sample z-test comparing IQ scores of students in gifted and talented program with general population
        one-tailed test, 76–78
        two-tailed test, 71–76
    social class and, 18
    z-scores, percentile rank and IQ scores, 59–62
Interpretation
    of correlation coefficient, 166, **166**
    in hypothesis testing, 73, 81, 96, 126, 143, 152, 186
    of IQ scores, 54
        Iowa Test of Basic Skills, 62
    of SAT scores, 54, 55
Interval scales, 10, 11, 19, 71
Iowa Tests of Basic Skills (ITBS), 17, 62, 63

L

Lehmann, Nicholas, 57
Leibniz, Gottfried, 2
Linear regression, 3

M

Mass measurements, 11
Matched pairs t-test. *See* Dependent t-test
Mean, 28, 30–31, 40
    analysis of variance for, 141–162
    calculation of, **30,** 30–31
        with graphing calculator, 45–46
    definition of, 29, 30
    of normal distribution, 30, **30**
    probable error of, 4
    regression to, 3
    standard deviations from, 35–39
        z-score and, 55
Measurement
    definition of, 10
    formative, 10
    summative, 10
Measurement scales, 10–11, 19
    interval, 10, 11, 19, 71
    nominal, 10, 11, 19
    ordinal, 10, 13, 19

INDEX

ratio, 10, 11, 19, 71
review questions on, 20
for temperature, 10, 11
Median, 28, 31–32, 40
calculation of, **31,** 31–32
definition of, 29, 31
of skewed distributions, 32
Mendel, Gregor, 10
Middle Ages, 2
Mode, 28, 40
bimodal distribution, 32
calculation of, 32, **32**
definition of, 29, 32
of skewed distributions, 32, **33**
Multivariate analysis, 4
Murray, Charles, 18

N

Negatively skewed distributions, 17–19, **19**
Newton, Isaac, 2
No Child Left Behind Act, 50
Nominal scales, 10, 11, 19
chi-square test of data from, 183–184, 188, 193
Nonparametric statistics, 183, 193
Normal distributions, 16–17, **17,** 19, 53, 55. *See also* Bell (normal) curve
areas of the standard normal distribution, 75, 207
of male height, 53
mass production and, 54
measures of central tendency in, 29, **30**
percentile ranks and, 53–54, **54**
standard deviation and, 35–40
z-scores and, 55–56, **56, 57**
Norm-referenced data, 35, 50, 53, 55, 64
Null hypothesis (H_0), 70, 72, 80, 95, 105, 126, 142, 151, 185
for analysis of variance, 143
for chi-square of independence, 189
for chi-square test of goodness of fit, 186
for dependent t-test
one-tailed test, 120, 123
two-tailed test, 116
for independent t-test
one-tailed test, 105, 109

two-tailed test, 110
for one-sample t-test
one-tailed test, 99, 101
two-tailed test, 96, 99
for one-sample z-test
one-tailed test, 76–77
two-tailed test, 73, 75
reject or do not reject, **71,** 73, 81, 96, 127, 143, 151, 185
type-I errors and, 70, **71**

O

Observed frequency, 184, 187, 190
Ordinal data, 51, **51**
calculation of percentile rank from, 52
Ordinal scales, 10, 13, 19

P

Parametric statistics, 71, 151, 193
Pearson, Karl, 3, 4, 94, 184
Pearson correlation coefficient, 3, 165–170, 177
computing with graphing calculator, 168–170
definition of, 166
formula for, 166–167
interpretation of, 166, **166**
numerical values of, 166
review questions on, 178–180
for SAT test scores, 166, **166, 167,** 167–168
Pearson product–moment correlation, 166
Percentile ranks, 50–52, 63–64
calculation of, 50–52
definition of, 50
normal distributions and, 53–54, **54**
z-scores and, 50, 54–59
IQ scores and, 59–62
Phi coefficient (Φ), 190
Plackett, R. L., 184
Positively skewed distributions, 17–19, **19**
Post-hoc tests, 4, 146–149
Probability, 2
Probable error of the mean, 4
Psychological tests, 11

INDEX

p-value, 70
Pythagoras, 2

R

Range, 34, 40
Ratio scales, 10, 11, 19, 71
Regression to the mean, 3
Related measures t-test. *See* Dependent t-test
Relationship
    chi-square test of goodness of fit for, 184–187
    correlation coefficient of, 166
    phi coefficient for, 190
    strength of, 166
Reliability, 165, 178
Rounding numbers, 19

S

Sample standard deviation, 35–39, **37–38**
SAT test scores, 10, 16
    interpretation of, 54, 55
    one-tailed t-test of impact of SAT prep program on, 99
    Pearson correlation coefficient for, 166, **166, 167,** 167–168
    percentile rank and, 50
    range of, 50
    recentering of, 59, **60**
    scatter plot of, 170, **171**
    socioeconomic status and, 57
    Spearman rank-order correlation coefficient for, 175, **176**
    as standard scores, 49, 54, 63
    T-score and, 63
    z-score and, 56–59
Scatter plots, 170–173, **171,** 177
    of high positive correlation, 172, **172**
    of low negative correlation, 173, **173**
    of no correlation, 173, **173**
    of perfect negative correlation, 172, **172**
    of perfect positive correlation, 171, **171**
Scheffe, Henry, 146
Scheffe's method, 146–149
Seven steps for hypothesis testing, 72–73, 80–81, 95–96, 126–127, 142–143, 151–152, 185–186
Significance test, 70, 81, 127
Simple bar graph, 14, **16,** 19
Skewed distributions, 17–19, **19**
    median of, 32
    mode of, 32, **33**
Spearman, Charles, 53
Spearman rank-order correlation coefficient, 175–177
    calculation of, 175–177
    definition of, 175
    formula for, 175
    numerical values of, 175
    review questions on, 179–180
    for SAT scores, 175–176, **176**
Standard deviation, 35–40
    calculation of, **36–37,** 36–39
        with graphing calculator, 45–46
    formula for, 38, **38**
    mean and, 35, 39, **40**
    normal curve and, 35, **36**
    with norm-referenced data, 35
    outliers and, 39
    in t-test, 94
    variance and, 36, 39, 40
    z-scores and, 55
Standard Error (SE)
    estimated, for t-test, 94, 98, 101, 104, 108, 113, 118–119, 122–123
    F-ratio for analysis of variance, 144–148
    for one-sample z-test, 75, 78
Standard scores, 49–67
    definition of, 54
    normal distributions and percentiles, 53–54, **54**
    percentile ranks, 50–52
    review questions on, 65–67
    in schools, 50
    T-scores, 50, 62–63
    z-scores and percentile ranks, 50, 54–59
        IQ scores and, 59–62
Stanford-Binet Scale, 17, 55
Statistical bias, 4
Statistical calculations, 5
    with graphing calculator (*See* Calculator explorations)

INDEX

Statistical significance, 70, 81, 127
Statistical test, 2, 6, 73, 81, 96, 126, 143, 151, 185
Statisticians, early, 2–5, 6
    Fisher, 4–5, 70, 81, 94, 127, 141
    Galton, 3, 53
    Gosset, 4, 93–94
    Pearson, 3, 4, 94, 184
Statistics
    applications of, 5, 10
    definition of, 2, 10
    history of, 2–7
    inferential, 3, 74, 80, 81, 127
    mathematics and, 2, 5
    nonparametric, 183, 193
    origin of term for, 2
    parametric, 71, 151, 193
Student's t-test, 93

T

t-distribution, 4, 94, 97, 100, 106–107, 117, 121, 138–139
Temperature measurement, 10, 11
The Bell Curve: Intelligence and Class Structure in American Life (Hernstein and Murray), 18
The Big Test (Lehmann), 57
The Grammar of Science (Pearson), 3
The Method (Archimedes), 2, 5
The Mismeasure of Man: The Definitive Refutation to the Argument of the Bell Curve (Gould), 18
Thorndike, E. L., 53
Time measurements, 11
Time reversal, 3
T-scores, 50, 64
t-test, 93–137
    analysis of variance for multiple t-tests, 141–146
    dependent (related measures, matched pairs), 115–125
        computing with graphing calculator, 123–125
        example of one-tailed test, 119–123
        example of two-tailed test, 115–119
        use of, 115

    formula for, 94, 98, 101
    independent, 103–115, 126
        computing with graphing calculator, 113–115
        example of one-tailed test, 105–109
        example of two-tailed test, 109–113
        formulas for, 104
        use of, 104
    one-sample, 94–103, 125
        computing with graphing calculator, 102–103
        example of one-tailed test, 99–102
        example of two-tailed test, 95–99
    origin of name for, 93–94
    review questions on, 127–137
    t-distribution, 4, 94, 97, 100, 106–107, 117, 121, 138–139
Type-I errors, 70
    alpha level to determine probability of, 72, 74, 80, 95, 97, 127
    analysis of variance to protect against, 141–142
    court room trial example of, 70, **71**
    definition of, 70, 74, 80
    hypothesis testing to minimize risk of, 81

V

Validity, 165, 170, 178
Variability, 28, 34, 40, 144
Variables, 11–12
    continuous, 11–12, **12**, 19
    correlation of relationship between, 166–180
    definition of, 11
    discrete, 11, 12, 19
    in educational testing and measurement, 11, **12**
    Pearson correlation coefficient for, 166–170
Variance(s), 28, 34, 35–40
    analysis of, 4, 6, 141–162
    pooled, 104, 108, 112–113
Variation, 34–40
    central tendency and, 28, 34
    practical uses for teachers, 28, **28**
    range, 34

INDEX

review questions on, 42–45
standard deviation and variance, 35–40
Venerable Bede, 2

W

Wechsler Intelligence Scale for Children (WISC III), 54
Wechsler IQ Test, 55, 63

X

x-axis, 13, **13**

Y

y-axis, 13, **13**

Z

z-scores, 50, 55–59, 63–64
  applications of, 55–56
  calculation of, 55
  definition of, 50, 55
  normal distribution and, 55–56, **56, 57**
  percentile ranks and, 50, 54–59
    IQ scores and, 59–62
  T-scores and, 62–63
z-test, 69–91
  alpha levels and critical values for, 70, 74–75, 77, 81
  application to research, 71
  calculation for, 75, 77–78
    with graphing calculator, 79–80
  court room trial example of, 70, **71**
  formula for, 75, 77
  one-sample, 71
    example of one-tailed test, 76–78
    example of two-tailed test, 71–76
  Pearson correlation coefficient and, 167, **167**
  review questions on, 81–91